MANAGEMENT AND ECONOMICS OF BUILDING PROJECTS

P. K. Guha, M. I. S., F. I. E. (India)

Chartered Engineer & Valuer,
Retired Executive Engineer,
Public Works Department,
Govt. of West Bengal,
Formerly Director,
Calcutta Municipal Corporation

NEW CENTRAL BOOK AGENCY (P) LTD.

8/1 CHINTAMONI DAS LANE, CALCUTTA 700 009 (INDIA)

MANAGEMENT AND ECONOMICS
OF
BUILDING PROJECTS

First Published : 1997

Typesetter, Printer and Publisher : New Central Book Agency (P) Ltd.
8/1 Chintamoni Das Lane, Calcutta 700 009

Project Team : Pradip Das and Subas Maitra

ISBN 81-7381-267-5

Price : **Rs. 75.00** (Rupees seventy-five) only

To the memory of
my mother
Late Sarojbala Guha

FOREWORD

Shri P. K. Guha, as I have known, is a highly dedicated professional. In pursuit of his profession, the expertise and knowledge he acquired over the years in the complex management of building projects and his academic bent of mind, is reflected in his book titled **'Management and Economics of Building Projects'.** His skilled handling of the elements in the book covers all aspects of management of building projects as encountered by practising Civil Engineers in the field, particularly in monitoring of building projects in respect of their duration and cost targets.

It is a useful handbook for all Civil Engineers including diploma-holders engaged in Building Projects. Written in lucid style, it can be handled and grasped by budding professionals in the building trade. The management techniques employed to economize on construction cost have also been amply illustrated for the engineers and students uninitiated in the technique adopted in practice. LOB techniques have seen elaborated with illustrations as applicable to components of mass production of repetitive elements of building projects. Comparative suitability of techniques e.g. Bar Chart, PERT and LOB are also narrated in the text. The concluding chapter, specially the Labour and Material Analysis presented by the author would be very useful appendage for the purpose of control, regulation and monitoring of resources in the building projects.

To sum up, the author has taken enough pains to present a Text Book which will enrich the professionals pursuing Building Projects of magnitude where economy to the tune of 5% may lead to large quantum of gain in the ultimate analysis. Shri P. K. Guha richly deserves congratulations for his works and encouragement for more such endeavours.

Calcutta
18th December, 1996

M. S. Maitra
Retired Chief Engineer,
Public Works Department, Govt. of West Bengal
Formerly Director General Operations,
Calcutta Metropolitan Development Authority

PREFACE

Civil engineering has become a part of the society since inception of the civilization and will continue to be so till it exists. With the development of civilization, demand for meeting the needs of the human being has increased. So the civil engineering has come forward with new innovations, analysis, design and execution-technology to meet the demand. Thus the bridges, roads, railways and various types of buildings have been built with challenging problems.

With demand for more construction, the materials available started becoming scarce, financial resource became stringent; but expectations of the people grew high. This necessitated the introduction of management technology for economising financial and material resources, better utilization of manpower and timely completion of the projects.

In Indian condition, the construction-technology is a complex one. The construction industry uses labour intensive methods, and thus takes the responsibility of employing large number of labour force, who mostly migrate from rural areas for gainful employment. Otherwise, by using machineries the project duration could be shortened and quality improved by eliminating human error; but unemployment would have been increased. Thus, the method prevailing now is a complex one as mechanization has been adopted in limited areas keeping the project more labour intensive. In such condition, management of the project assumes greater importance and cannot be controlled without adoption of any methodical system in all the aspects of project management.

We have large number of architects and engineers, both of graduate and diploma level. Institutions impart specialised construction management techniques PERT, CPM, Network systems including use of computers; but very little has been done for training the human resources directly responsible for implementation of the projects. This gap has been created due to misconception that Management is a subject to be studied and may be applied in large industries involving high-tech methods and systems of production. So far as civil engineering projects are concerned, specially the smaller ones and the building projects, management depending mainly upon previous experience was considered enough and the same was again based on the knowledge of the masons, carpenters, barbenders, welders, plumbers, electricians, concrete work, force, etc. Most of the building construction trades have been passing from generation to generation without upgradation of the knowledge and skill. The management technique of the construction has remained stagnant.

In this book, attempt has been made to bridge the gap by trying to make the construction management technique easily understandable by all concerned in the industry from the higher to lower level of the management.

The discourse on the subject has been limited for application in cases of building construction only as buildings perhaps constitute the largest body of national assets created both at the public and private expenses. As in all projects be it a civil engineering project or belong to other discipline, building always occupies a significant portion. With minor modifications, the discourse can be utilised for management of construction projects other than buildings.

The subject is a new one and is still developing. Fundamentals of Management, PERT, and CPM have been explained. Integrated management system, planning for mass housing have been discussed with illustrative example. There would still be scope for further improvement in presentation of the methods and ways of application. There may be shortcomings and lapses in the discussions. It would be highly appreciated and gratefully acknowledged if any suggestion and or correction is forwarded.

I am grateful to Mr. Amitabha Sen, Director, New Central Book Agency (P) Ltd. for bringing out the book in very short time. I am deeply indebted to Shree Shyambenode Banerjee and Shree Nandadulal Banerjee who extended all help in writing the book. I an thankful to Shree Dilip Kumar Das who took the trouble of typing the manuscript. I also thank the authors of the books and papers mentioned under the 'Reference' for the enormous help I received from the knowledgable information imparted by them in their writing.

Anandalok—A/8
A/1/1, C.I.T, Scheme—VII M
Calcutta 700 054
August, 15th, 1996

P. K. GUHA

CONTENTS

LIST OF ILLUSTRATIONS AND GRAPHS

LIST OF TABLES

Chapter 1

Construction Management

1.1.0 INTRODUCTION

In a developing country, volume of construction activity goes on increasing. So has construction component increased manyfold in recent times. It would be evident from Table 1.1. below that investment on construction component is necessarily very high.

TABLE 1.1

Approximate construction components in sectors

	Sectors	Construction component as per percentage outlay
i.	Agriculture and allied sectors	34 %
ii.	Irrigation and flood control	60 %
iii.	Power	75 %
iv.	Industries (Organised)	30 %
v.	Transport and communication	40 %
vi.	Education	20 %
vii.	Health	37 %
viii.	Housing and Urban development	100 %
ix.	Social welfare	21 %
x.	Labour welfare	45 %

In the Sixth Plan, the allocation for construction component was approximately 36.2% which has increased to 47% in the Seventh Plan. Again, in the construction activities, the share of Building is the largest as, whatever project is undertaken, construction of building is a must.

The primary goal of a project is a quality product within budget, on time and with little or no litigation. To achieve such goal it is essential that the project is managed properly and effectively. Due to increase in volume of building construction activity, its diversity and input of modern techniques and innovations, the traditional concept of project management is becoming obsolete.

This is becoming more difficult on account of rising expectation of the decision makers and the beneficiaries, the people to complete the construction in tight time schedule. All these facts of the day call for adoption of available modern management techniques in the field of Building Construction.

1.2.0 WHAT IS MANAGEMENT

During construction process many difficult and complex management problems generally arise. To overcome them and to ensure judicious allocation and efficient use of resources, improved management techniques need be adopted. The objective of the management is to make the best use of all the available resources. One of the most serious problems faced by construction industry is related to time and consequent cost over-runs. Modern management techniques available are vital tools for tackling these problems and ensure timely and economical completion of projects.

MANAGEMENT

Management is a scientific tool for organising the methodical approach of planning, programming and execution of a project and also monitoring the progress of the work including quality control, scheduling of materials, labours and plants and machineries that would be required for the project.

A project is composed of various activities and they are interrelated. The activities or functions need be coordinated and monitored for timely completion with quality control so that the total project is completed in the time frame specified. During this process, other functions also get into the process.

i. **Material management**

Materials of various categories and in large quantities are required for execution of the project. These need be procured timely so that there may not be any delay of the item due to non-availability or shortage of materials or any particular material.

ii. **Labour management**

Different categories of labours are required at different stages of construction. This also has to be looked before and planned properly so that there may not be any hindrance due to shortage of any category of labour at any stage.

iii. **Plants and Equipments**

Various plants and equipments (machineries) may be needed at different stages of construction project. These need be planned properly so that they are available at the right time. Any of the plants and machineries may break-down during progress of work. Suitable standby should be kept so that the same may be put into service to keep the execution schedule as per programme and maintain its normal progress.

1.2.1 OBJECTIVES OF PROJECT MANAGEMENT

Project management is a highly specialised job. It becomes more complicated as the process involves human factor at all stages. The involvement of human factor need be taken into account to achieve the objective of the management.

The objectives are :

i. Completion of the project allowing for minimum duration of time,

ii. The project should be completed without delay using the available manpower and other resources,

iii. The project should be completed without any delay and with minimum capital investment and without compromising with the quality.

1.3.0 PHASES OF PROJECT MANAGEMENT

Project management comprises of three phases which are to be taken care of:

i. Planning and programming

The project should be planned properly and the different activities programmed according to sequence of operation.

ii. Scheduling

Scheduling of the activities, scheduling of the materials required and scheduling of the plants and machineries that would be necessary.

iii. Control

Project controlling or monitoring i.e. keeping observation on the performance so that no overrunning of time occurs.

1.4.0 PROJECT PLANNING AND PROGRAMMING

On conceiving a project, the concerned Engineering department need be intimated of the relevant requirements to enable them to prepare preliminary plans and estimates for obtaining administrative approval of the project. Study of the requirements need be made for achieving optimum construction costs. Selection of site and acquisition of land and its development also need be started simultaneously. These take considerable time and necessitate proper planning and phasing of the work. Programming of all the operations for execution of the work is essential and should be done with all possible care. The various operations in a construction programme are to be so synchronised in such a manner that the expenditure incurred at each stage does not remain idle for long. The programme should include the quickest and most economical methods of execution of work, consistent with the different operations of the construction. The programming should ensure continuous productive work in optimum time.

The planning part of management is followed by programming.

a. Planning

Planning of a project means formulation of the project and sanctioning of it by the appropriate authorities. The procedure comprises essentially of the following activities.

i. Investigation for preparing the preliminary project report,

ii. Preparation of preliminary project report,

iii. Administrative approval of the project by the concerned Administrative department,

iv. Initiation of the land acquisition proposal,

v. Investigation for the detailed project report,

vi. Preparation of the detailed plans and estimates and preparation of the detailed project report,

vii. Technical sanction of the project by the competent authority in the technical department to whom the job has been assigned,

viii. Budget allotment,

ix. Tendering procedure,

x. Acceptance of the tender,

xi. Supervision and execution of the work,

xii. Accounting work, and

xiii. Maintenance of the works in operation.

i. Investigation for preparation of the preliminary report

On the basis of the request and requirement of the client or the department, information are to be collected for preparation of the project and a preliminary or a rough cost estimate. Although detailed investigations are not undertaken at this stage as it may involve heavy cost, it should be taken care of that on the basis of the preliminary investigation, preliminary project report with cost estimate would be prepared and administrative approval would be sought for on it. The preliminary cost estimate should not fall far off from the detailed cost estimate necessitating revised administrative approval.

ii. Preparation of preliminary project report

Preliminary project report with preliminary or rough cost estimate shall be prepared furnishing an idea and the financial aspects of the project. The client or the department after study of the project report will ask for next stage work, i.e. a detailed project report. Administrative approval of the project may sometimes be accorded on the basis of preliminary report, but rationally it should be accorded on the basis of the detailed report.

iii. Administrative approval by the concerned Administrative Department of the client

The concerned department or the client on receipt of the project report will go through it and examine the information and analysis furnished, examine the physical and financial viability and specially the availability of the fund in the department. They may require clarification on various points and the report may have to be modified or adjusted to suit the desire and interest of the client. Obviously, the project belongs to them. When the project report is found acceptable and the fund or the source of it is available, the concerned department or the client shall accord Administrative approval of the Project.

Administrative approval of project is normally accorded on the basis of detailed project report supported by a detailed cost estimate. Preparation of detailed project report and cost estimate require considerable time. For saving time and expediting the project, Administrative approval is often accorded on the basis of preliminary project report and rough cost estimate.

If the project is proposed to be funded by taking loan from any financial institution in stages for a fixed period with a rate of interest, the same have to be taken into consideration in the financial analysis. Dialogue may have to be made with the institution to convince them regarding repayment of the loan.

iv. Initiation of land acquisition proposal

On conceiving a project, proposal for acquisition of land required for the project should be initiated without any delay. The process of acquisition of land is a long-drawn one and in most cases causes inordinate delay due to various litigations etc., resulting not only delay in completion of the project but increasing the cost of the project. Land acquisition proceedings should be initiated along with other actions regarding planning of the project so that possession of the land is available as soon as the project is sanctioned.

v. Investigation for the detailed project report

On receipt of the approval of the preliminary report, detailed survey of the site is undertaken, detailed investigation and studies are undertaken for preparation of the detailed project report. The investigation and studies shall cover wide range – analysis of the requirement and finalisation of the provisions, study of soil, study of environment, ecology, transport, communication, weather specially rainfall, seismic effect, flood levels etc. All possible information shall be collected and incorporated in the report. The report shall consist of an index plan with reference to some important object nearby so that the location may be identified without any difficulty. The report should indicate the distances of the important places and public utility services and a cost estimate based on unit area basis – rates based on similar project executed currently in the area or obtained by analysing it. A cost benefit ratio shall be shown justifying the economic viability of the project.

A report on physical viability of the project shall also be incorporated. A report on amortisation of the cost of the project together with a cash flow chart showing the in-flow and out-flow of fund shall be furnished. In conclusion, a report shall be incorporated justifying why the project should be taken up and what would happen if the same was not taken up.

The detailed project report shall cover all possible technical, financial and social information and remarks and their implication on the project. All possible analysis of the information shall be incorporated. For example, if it is a housing project, calculation of requirement of water, power and sewage disposal shall be included. The areas provided for dwelling or other purpose shall be justified. In the case of any road project involving large volume of traffic, the road width and depth of hard crust provided shall be justified by proper and acceptable calculations.

Detailed information regarding availability of construction materials like bricks, stones, stone chips, sand, timber and specially water need be collected and supplied to the planning and estimating section of the project department. Information regarding availability of local labours both skilled and unskilled with their wages need be collected. A regular flow of these information is desired by the Planning and design branch for preparation of an authentic document, 'Project Report' and Detailed Estimate.

vi. Preparation of detailed plans and estimates

When the project report has been approved by the concerned Administrative department and the same has been communicated, detailed plans and estimate of the project shall be prepared. This shall be preceeded by design of the project and its component parts. While designing, rapport must be kept with the concerned department who will be in charge of execution, regarding the design matters and technical matters, so that there may not be any dispute at any later stage. This will have to be done through discussions time to time which will eliminate controversies and save much time.

All references and authorities and code shall be furnished for ready reference of the client's technical department regarding design and construction technology.

Preparation of complete detailed plans and estimates may take substantial time. However, some of the detailing and construction drawings which will not affect the cost may be left for furnishing later and a cost estimate may be drawn. The rates of the cost estimate shall be adopted from a standard 'schedule of rates' with consent of the client. The cost estimate shall be prepared carefully and as far as correctly. If the cost of construction varies more than 5% of the estimated cost, a revised estimate shall have to be prepared for Administrative and Technical sanction of the concerned authorities. To ensure guard against increase in cost due to price contingency, provision for escallation of cost year-wise shall be made. This shall be made in accordance to the R. B. I. price indices for both labour and materials as available and then projecting the same for the proposed construction period.

Materials and Labour Requirements

On completion of estimate of work, an estimate of manpower and material requirements is generally required to be prepared. This is required for various purposes such as budgeting and procurement of materials, specially the scarce materials, computation of cost indices, justification of tenders etc. The estimation may be done in various ways :

a. Approximate requirements may be obtained on percentage basis of the cost of the structure. This is a very approximate method. Tables for approximate estimation of materials and labour are generally maintained by the estimating branch.

b. The most dependable method is to compute the requirements from the quantities of the detailed estimate. The computation can be readily done on the basis of charts and tables available in all the 'Schedule of Rates'.

c. C. B. R. I. has established quicker methods based on Statistical equations which relate computation of materials/labour with the plinth area of buildings.

vii. **Technical sanction of the Project with cost estimate**

When all the above procedures have been complied with and the technical department of the client is satisfied and the cost estimate is within 5% of the preliminary detailed cost estimate on which administrative approval had been accorded, the appropriate authority in the client's technical department shall accord technical sanction of the project. If the detailed estimate varies more than 5% of the preliminary cost estimate, revised Administrative Approval would be required.

The technical sanction together with a set of plans and estimate shall be forwarded to all concerned including the department entrusted for the execution.

viii. **Budget allotment**

Technical Department entrusted for execution shall wait for provision of the full amount of the project or part in the Budget estimate of the year. Generally, token provision is kept in the budget in anticipation to avoid delay. When the detailed cost estimate is ready and the total amount with the duration of construction is known, appropriate amount is provided in the revised budget estimate generally at the later part of the financial year.

ix. **Tendering procedure**

After technical sanction of the project preceeded by the Administrative Approval and having provision of appropriate amount in the Budget, the department entrusted for execution of the project is entitled to commence execution after inviting tenders in proper method, processing them and finalising the tender found suitable. This part of the procedure is also time consuming. Generally, some actions which can be standardised are taken beforehand. Similar other actions regarding the particular project which are possible to be taken in anticipation are also taken to curtail delay.

Preparation of tender documents

Tender documents generally comprises of four/five parts :

Part I — Contains description of the work, general rules and directions, standard terms and conditions of contract, schedule of materials with prices proposed to be issued by the department.

Part II — Contains standard technical specifications of materials and workmanship.

Part III — Contains schedule of items of works with quantity, unit, estimated rates and amount.

Part IV — Contains complete set of drawings furnishing details of work as far as possible. These drawings are marked 'meant for tendering and not for execution'.

Part V — Apart from the above four parts, in some cases another part is added which contains technical specification of any special item or items of work which would require any particular technology or materials or any special tools and plants.

Parts I and II are standard and are always kept ready in printed form for saving time and are specified by form number. Parts III, IV and V are prepared as required for each individual tender separately and are added to the document.

All these parts make the tender documents complete. Complete set of documents is bound in book form and is priced for sale to intending tenderers.

Tender Notice

Notices are issued to various departments and organisations and also released through newspapers for wide publicity of the tender. A specimen of tender notice is given below. This tender notice is generally termed as short tender notice and the documents complete with all the parts is termed as detailed notice.

Specimen Tender Notice

NOTICE

Office of the ....

Tender number ....

Sealed tenders are invited from the experienced, bona fide and resourceful contractors of appropriate class of enlisted contractors and also from the outside reputed contractors for the under mentioned work :

1.	Name of the work	:	Here title of the project and the part for which the tender has been called for is to be mentioned.
2.	Short description of the work	:	Brief description of the work intended to be executed is to be mentioned.
3.	Estimated cost of the work	:	Rs.
4.	Earnest Money	:	As per terms and conditions.
5.	Time for completion of the work	:	As desired.
6.	Manner of depositing earnest money	:	As mentioned in the terms and conditions.
7.	Source of receiving tender documents	:	As suitable and in practice.
8.	Person to whom the tenders are to be deposited	:	As desired.

Any other information in connection with the tender may be had from the .... on all working days during office hours up to the date of tendering.

....

Person inviting tender

Quoting of rates by the tenderers or offering tender rates :

There are different methods of tendering :

a. **Two parts tender**

Tenders may be invited in two parts, Part I would contain all technical matters without any indication about the rate, the contractor proposes to quote. Sometimes, in this part contractors are asked to submit alternative design of structure, if they liked. Contractors are asked to submit their terms and conditions, if they so desired.

In the second part, contractors are asked to offer their proposed rates or the financial proposal.

The first part of the tender is opened in presence of the tenderers who have submitted tenders and desired to be present. The tenders are then scrutinized, and analysed after getting clarifications. The tenders which are found to be technically sound and acceptable are selected and their other part—i.e. the second part or the financial part—is opened in presence of the selected contractors. Their rates are disclosed. These are then scrutinized and evaluated and finally the tender which was technically sound and financially favourable is selected for acceptance.

This method of tendering is adopted for very large and complicated works.

b. **Item rate tender**

In cases of tenders of moderate value, the general method is to furnish the estimated rates and amount in the schedule of items or 'bill of quantities' and the contractors are asked to quote their contractual percentage on the schedule which shall be applicable on all the items.

The method has been found to be suitable for tenders of moderate value and having no complication.

Submission of tenders

Tender documents complete in all respects and accompanied by the evidence of deposit of requisite earnest money are submitted by the tenderers within the specified date and time to the officer calling for the tender or to the officer as mentioned in the notice.

x. **Acceptance of the tender**

After scrutiny and evaluation, when a tender has been selected for acceptance, the matter is conveyed to the contractor with the request to deposit the balance earnest money and submit the formal tender for formal acceptance of it or is asked to enter into an agreement.

In the public works system, the acceptance of the selected tender on behalf of the Governor of the State is considered as agreement. On receipt of the letter of selection of the tender, the tenderer has to deposit the balance security deposit and submit additional copies of tender for duplicate and triplicate copies of the accepted tender. In cases of statutory bodies a formal agreement is entered into by the parties on stamp paper enclosing all the papers of the tender documents and other clarifications or any other terms agreed upon by both the parties.

xi. **Supervision and execution of the work**

On receipt of the work order, the contractor commences the work and proceed with it. The responsibility of the department is to exert proper supervision to ensure the quality control of the work so that no dispute arises, specially for the portions which would be concealed or covered up. Although the primary responsibility of the contractor is to ensure the quality by exercising strict supervision, still the responsibility of the department or the client is no less as ultimately, he is the owner of the structure and shall have to pay for the work on being satisfied with the quality and quantity. Department or the client shall have to check the work and on being satisfied, his representative would measure up the work and record the same in the Measurement book for payment.

In case of large works, one or more representative of the client should be stationed at site. They shall check the materials collected, whether those conform to specification and as per sample submitted by the contractor and approved by the authority. The quality of the work at all stages shall be controlled by testing the works done and checking the lines and levels. If any material collected or work done is found to be under specified, the contractor should be asked to remove the material from the site forthwith and the under specified work shall be asked to demolish.

xii. **Accounting work**

With the progress of work, the quantities of the items shall be measured and recorded and interim payments made. On completion of the work, the representative of the client shall record final measurement and being satisfied, would arrange for final payment. After final payment of all works, proper accounting of the works, both for quantities of each item and cost thereof shall be made. Accounts of materials issued by the department and cost thereof already realised shall be finalised. When all these accounts are finalised, a clear picture regarding all the aspects of the work would be clear.

xiii. **Maintenance**

On completion of the construction and when the structure is in use, maintenance and operation would start to keep the structure useworthy during its life cycle and to maintain the supply of essential items necessary for using the structure e.g. water supply, power supply etc.

Occasional and regular repairs and replacement of parts of the structure under maintenance would also be required for proper upkeeping of the structure and for prevention against early decay of the members of the structure.

1.5.0 SYSTEM FOR CARRYING OUT CONTRACT

The recognised systems for carrying out work other than by the employment of daily labour are, 'piece work' and 'contract work'. Piece work is that for which only rate is agreed upon without reference to the total quantity of work to be done in a given period.

All other works done under agreement is termed 'contract work' and in agreements for such work, which should invariably be in writing, there should generally be a stipulation as to the quantity of work to be done, and the time within which it is to be completed.

1.6.0 CONTRACT

Contract is an agreement between two persons declaring their common intention to do or not to do an act. All agreements are contracts, if they are made with free consent of the parties competent to make contract for lawful consideration with a lawful object as defined under Section 2 (k) of Indian Contract Act of 1872.

1.6.1 BASIS OF CONTRACT

Variation of cost is expected, according to the basis on which a contract is placed, and this must be taken into account when estimating and decided upon the form of contract. Studies carried out by the CBRI has shown that lump sum contract performs the best in respect of completion time, completion cost and time and cost overruns for repetitive type of construction. Competitive tenders based on bill of quantities is also likely to produce keen competition, resulting reduction in cost. Negotiated contract on a 'bill of quantities' will not be economical when compared with the lowest tender as obtained in a competitive method. Sometimes 'cost plus' contract is entered into, where actual cost plus an agreed percentage is paid to the contractor. A higher cost of work will give a higher amount to the contractor. Consequently, the cost of construction is likely to be higher in this type of contract.

1.6.2 TIME FACTOR

The cost of a project depends on the completion time. Overhead costs of a contractor per unit time is usually constant. Delay in the completion of the project delays return on the investment, consequently results in increased cost to the contractor and to the owner. Optimum completion time for a project depends upon the quantum of work, infrastructural facilities and availability of materials and labour. When a project is asked to be completed before the optimum completion time, overtime working at enhanced overtime rates has to be adopted. This would increase the cost of the project.

1.6.3 INGREDIENTS OF CONTRACT

An agreement should contain the following :

i. Proposal.

ii. Communication of proposal and its acceptances.

iii. Consideration.

iv. Consent.

v. Competency to contract.

vi. Communication when complete, and

vii. Revocation of proposal and acceptance.

1.6.4 CONTRACT DOCUMENTS

Contract document consists of

i. A complete set of drawings with details explaining the work

ii. Complete specifications of items of works, including any special specification for any item

iii. Schedule of quantities of various items of works

iv. Complete set of terms and conditions including any special terms and conditions, and

v. All correspondences and clarifications made and agreed upon.

Contracts may be of three kinds :

(a) Lumpsum contract — contractor is engaged to execute the work with all its contingencies for a fixed sum.

(b) Schedule contract or item rate contracts are those in which the contractor undertakes to execute the work at fixed rates; the sum he is to receive depending upon quantities and kind of work done or material supplied. In this system the contractor has the advantage to manipulate the rates of items. He can quote higher rates for the items to be executed at the early stage and compensate the same by quoting lower rates for items to be executed at the later stage – thus keeping the total amount competitive.

(c) Percentage contract

It is identical in all respects with the item rate contract. In this system, the 'bill of quantities' with rates and amount is given. The contractor is asked to quote his percentage on the schedule given, which would be applicable on all the items. Thus the drawback of the item rate contract is eliminated.

1.6.5 COMPARATIVE PERFORMANCE OF DIFFERENT CONTRACT SYSTEMS

As the system of contract affects the completion time and cost of the project to a great extent, study of the comparative performance of various systems is needed for effective economy of the work. Lumpsum contracts are based on a fixed price including profit margin of the contractor and is agreed upon in advance and contractor has to complete the work in accordance with the plans, specifications, construction details and other contract documents which are supplied to him in advance.

In item rate contract and percentage contract, 'bill of quantities' is prepared for all the anticipated items. In item rate contract, items are priced by the contractor and payments are made for the amount of work actually executed. In percentage rate contract the contractor quotes on percentage basis on the rates provided in the 'bill of quantities'. The percentage quoted is applicable on all the items. Payment is made itemwise on the basis of quantities of work actually executed.

1.6.6 MERITS AND DEMERITS OF DIFFERENT CONTRACT SYSTEMS

Lumpsum contract is preferred for traditional and repetitive type of construction. The nature and extent of work is indicated on the drawings and specifications, with or without 'bill of quantities'. The significant attributes of this type of contract system are :

i. Commitments are known in advance;

ii. Better and easier cost control can be exercised by the contractor as he has more independence in the matter.

iii. Reduction in administrative work of the owner/client is possible during construction period resulting in better supervision as, much of the works are done by the contractor.

iv. Higher efficiency and speed in construction can be effected by the contractor leading to his increased profit.

v. Preparation of plans, construction details and specifications are possible in advance as, in general, these are done by the contractor.

vi. Contractor can quote better rates based on new techniques and/or innovation.

Works which entail large estimates using relatively few methods of construction and where the volume of work cannot be exactly determined in advance, item rate or percentage rate contracts are used. The significant attributes of these systems are :

i. A reasonably accurate estimate is a prerequisite;

ii. The system provides for more variations;

iii. Complete drawings and specifications are generally not required in advance. These are generally prepared and supplied with the progress of the work;

iv. Possibility of over–payment to contractor due to computation and book keeping is reduced;

v. Overrunning of time and cost due to delayed decision on drawings, details and specifications is a common phenomenon.

1.6.7 PERFORMANCE OF DIFFERENT CONTRACT SYSTEMS (STATISTICAL BASIS)

Statistical relationships of performance of different contract systems with regard to time and cost and total plinth area of buildings of projects were established. Comparison of the systems was done with respect to the following parameters :

i. **Completion time**

Lumpsum contract – least time.

Item rate contract – 45 – 75 per cent more time.

Percentage rate contract – 45 – 55 per cent more time.

ii. **Completion cost**

Apart from the normal variations like specification, number of storey, execution sites, design features, construction year, and degree of tender competition or bidding, volume of work influence greatly the completion cost in different contract systems.

Buildings with plinth area up to 1,870 sqm, performance of percentage rate contract is comparatively better.

Buildings with plinth area above 1,870 sqm., performance of lumpsum contract is better than percentage rate of contract.

iii. **Variation in time**

Projects completed within variation of time up to 50% from the stipulated time;

Lumpsum contract	—	79%
Percentage rate contract	—	36%
Item rate contract	—	30%

In rest of the projects the time variation was more than 50%. The comparative performance of lumpsum contracts was better.

iv. **Variation in cost**

Variation on completion remaining within 5% of contract cost ;

Lumpsum contract	—	85%
Percentage rate contract	—	46%
Item rate contract	—	45%

v. **Disputes**

Incidence of disputes leading to arbitration;

Lumsum contract	—	10%
Percentage rate contract	—	47%
Item rate contract	—	19%

Performance of lumpsum contract is better than percentage rate contract and item rate contract for projects like housing. Difference in the performance of percentage rate contract and its rate contract is very little.

1.7.0 SOME STANDARD TERMS AND CONDITIONS WHICH NEED BE INCORPORATED IN THE TENDER DOCUMENT PART I

i. **Earnest Money**

Earnest money is the amount the contractors are asked to deposit prior to tendering, demonstrating their earnestness for the tender. This is generally 1% – 2% of the estimated value of the tender and is asked to deposit one day prior to tender date. The earnest money of the unsuccessful tenders are refunded without any interest.

ii. **Security Deposit**

Security deposit is the amount which the tenderer, whose tender is selected for acceptance, is asked to deposit before formal acceptance of the tender or execution of agreement. This amount varies from 5% to 8% of the value of the tender, after consideration of the contractual percentages inclusive of the amount already deposited as earnest money. This security deposit is taken in two parts – one part about 2% including the earnest money is asked to deposit prior to acceptance of the contract or execution of agreement and the balance is recovered from progressive bills proportionately.

iii. **Supplementary items**

The schedule of items or bill of quantities shows the details of items of work to be executed. Any item of work not appearing in the schedule but required to be executed by the contractor under the contract is called supplementary item. Procedure for finalisation of the rate of supplementary item of work should be indicated.

iv. **Compensation**

When a contractor fails to complete the work within the specified period for no fault of client, he is liable to pay compensation for loss and damages suffered by the client. Procedure for realisation of compensation for the loss should be specified.

v. **Liquidated damages**

Liquidated damage is an amount of compensation payable by a contractor to the owner due to delayed construction having no relationship with real damage. If the contractor fails to complete the work within the time prescribed in the tender, then the contractor shall pay to the owner the sum stated in the tender as liquidated damages for such default and not as a penalty for everyday for the excess period taken between the date of completion specified in the tender or the extended time—as the case may be—and the date of actual completion of the work. The payment or deduction of such damages shall not relieve the contractor from his obligations and liabilities under the contract. If, before the completion of the whole work, any part of the work is occupied or used by the owner and is duly certified by the Engineer-In-Charge, the liquidated damages for delay shall be reduced in the proportion in which the value of the part so certified bears to the value of the whole of the works. The security money deposited and deducted from the contractor is held in full or part from the contractor during this period.

vi. **Extension of time**

The contract is made for execution of the work within a specified period. When the work is not complete within the period and the period of work is extended on agreement of both sides, the extension of the period is called extension of time.

vii. **Escalation of Prices**

Due to wide variation of prices of materials and labour in large work, provision need be made for price contingency of materials and labour. There are various formulae for evaluation of amount on account of price variation. The most acceptable formula is based on R. B. I. price indices for materials and labour.

Formula for price escalation/de-escalation

(a) **Materials :**

$$Vm = (0.60\ W - Md) \times \frac{Im - Im_o}{Im_o}$$

Where, Vm = Escalation/de-escalation on account of variation of prices of materials.

W = Total gross value of the work done at tendered rate.

Md = Value of the materials issued by the department at fixed rate.

Im = Wholesale price index number for all commodities as ruling during the period under consideration as published in the RBI Bulletin.

Im_o = Wholesale price index number for all materials other than issued by the department ruling at the time of tendering and as published in the RBI Bulletin.

(b) **Labour**

$$V_L = 0.25\ W \times \frac{I_L - I_{Lo}}{I_{Lo}}$$

Where, V_L = Escalation/de-escalation on account of variation of price of labour.

W = Total gross value of the work at the tendered rate.

I_L = Consumer price index number for industrial labour applicable for the region of the site of work as ruling during the period under consideration as published in R.B.I. Bulletin.

I_{Lo} = Consumer price index number for industrial labour applicable for the region of the site of work as ruling in the month of submitting tender as published in the R. B. I. Bulletin.

Procedure for computation of price escalation shall be indicated in clear terms.

viii. **Termination of the contract**

When a contractor fails to fulfil the terms of the contract and the work is incomplete, the contract may be terminated. Prior to termination, the contractor shall be served with a show cause notice asking him to explain why he would not be made liable to pay compensation for non-fulfilment of the contract. On receiving his reply and after due consideration, the contract may be terminated by the authority spelling out the mode of payment or realisation of the compensation. Procedure for termination of contract where such situation arises shall be delineated clearly.

ix. **Defects liability period**

After the work is complete, the contractor remains responsible for making good to any defects observed in the work for a specified period. This period is called defects liability period. The security money deposited or deducted from the contractor is held up in full or part during this period.

x. **Completion certificate**

On completion of the work in all respects including removal of all surplus materials, contractor's godown etc., clearance of site and being satisfied that there is no liability due from the contractor, the Engineer-in-Charge would inspect the work and issue a completion certificate. This should be recorded in the measurement book.

xi. **Refund of Security Deposit**

There are two systems of refund of security deposit :

(a) The total security deposit lying at credit of the contractor is refunded in full on recommendation of the Engineer-in-Charge and on certified by him that there was no liability due from the contractor after expiry of the defect liability period or maintenance period.

(b) The security deposit is sometimes refunded in two stages. 50% of the total security is refunded along with the final bill and the balance 50% after expiry of the defect liability period or the maintenance period. In both the cases the Engineer-in-Charge certifies that the work has been completed satisfactorily and there is no liability due from the contractor.

xii. **Ceiling on the Security Deposit**

Ceiling on the total deposit has been introduced. The quantum of ceiling varies in different organisations from Rs. 2.00 lac to Rs. 5.00 lac. Recovery of the part security from the progressive bill would be effected till the ceiling amount is reached.

xiii. **Work Order**

On completion of the contract or on formal acceptance of the tender, the Engineer-in-Charge issues order to commence the work allowing a short period for preparation, mobilisation and any other preparation.

xiv. **Mobilization period**

Certain period of time according to the nature and value of work is allowed before reckoning the date of commencement of the work. A clause mentioning the period of mobilization should be included in the tender document. The Engineer-in-Charge mentions the same again in the work order. Actual period of work is reckoned after the date of mobilization.

xv. **Work Programme**

During the mobilization period, the contractor prepares a work programme and schedule of materials that would be required to be issued by the department. The programme is finalised after mutual discussion and a Bar-Chart or a CPM network is drawn and agreed upon by both. The programme is followed throughout the period of work. The contractor is bound by the time limit i.e. the total period of work. He is also bound to produce proportionate progress of the work in proportionate time. He is also bound to follow the programme. The department is also bound to deliver the site of work immediately after the work order and extend all assistance in pursuance of the terms and conditions of the contract, specially, the issue of materials of good quality as and when indented by the contractor as per programme.

Chapter 2

Programming

2.1.0. PROGRAMMING

Planning & programming part of the management are the activities of the project effected prior to commencement of the actual work which are required for proper implementation of a project. Planning & programming are, therefore, essential prerequisites.

Programming of a project is most important for its proper implementation. Management of a project becomes difficult if the programming of it is not done carefully and consideration of its different components were not taken into account. Estimation of times that would be required for implementing the same also cannot be made properly. Mention has been made regarding work programme under contract which is submitted by the contractor in consultation with the client, but programming for the whole project including preliminary and planning activities need be made by the client or his appointed agent.

There are various methods of programming of a work; however, Bar-Chart and CPM are most effective methods and are widely applied.

Programming by Bar-Chart not only helps in proper and systematic approach of the project, but it helps in exercising control over the work during its implementation. Bar-Chart may be used in exercising control of the financial progress also.

2.2.0 SELECTION OF MANAGEMENT TECHNIQUE

The primary objectives of construction management are to utilise human, financial and material resources usefully and effectively. With the growing complexity and increasing volume of construction activities, traditional methods of construction management have failed to deliver the desired results. Improved, modern and effective method of management is needed for achieving the objectives. Bar-Chart, PERT, CPM, Line of balance (LOB), Integrated Management System (IMS) as developed by the CBRI are the techniques or tools for effective and efficient management control. The systems have their advantages and draw – backs. Selection of a particular system would depend upon the nature of work, whether indoor factory line production or outdoor field construction, specific purpose and suitability. Management by these modern tools require adequate training in the field. This also would be a factor for selection of a particular method.

Applying a technique without having adequate knowledge of the subject may lead to disaster instead of offering any potential help. After carefully judging the merit of the method, and its possibility of proper implementation, a particular method shall have to be adopted. All the methods would not serve the same purpose.

Apart from the methods as mentioned, there are other methods also. The latest trend is to introduce management control by computer. In the coming days some other better and precise methods would be available.

Whatever may be the method, the basic aims of management is to optimise the utilisation of the resources available and thus to achieve economy in all respects. When applied properly, management technique is sure to yield favourable results and prove saving of resources. Thus it would prove economical not only to the owner/client but also would save national resources.

2.3.0 BAR-CHART

In Bar-Chart method, the project is divided into a number of clear activities which can be identified separately. Each activity denote a complete item of work. A Bar-Chart is a chart showing the progress of different activities in respect of time. The duration of activities can be read from a Bar-Chart. These activities are presented linearly against vertical axis of a two co-ordinate axes diagram. The horizontal axes represent the duration period of activities i.e. time required for completion of the activities—which is estimated from the past experience of similar work. Each bar represents one specific item of work or activity of the project. The beginning and the end of each bar represent the time of commencement and the time of finishing of the activity. The lengths of the bar, therefore, represents the duration period of the activity.

As the Bar-Chart is represented in two axes—time against activity—probable expenditure and actual financial progress can also be shown on the chart with the help of the two axes—the horizontal axes indicating the time and the vertical axes showing the planned expenditure. The line on the chart obtained from the two co-ordinates generally takes S-shape and is, therefore, termed S-curve. Both planned and actual expenditures can be shown and curves obtained. The curves may be shown in different colours or one in firm line and the other in dotted line.

Thus a Bar-Chart can represent both physical and financial planning and progress.

Example

Activities		Duration
1.	Earthwork in excavation in foundation	2 weeks
2.	Foundation concrete and Brickwork up to plinth	5 "
3.	Brickwork in superstructure	6 "

4.	Shuttering and centering for roof slab	4	weeks
5.	Laying of reinforcement	2	"
6.	Concreting of roof slab	4	"
7.	Releasing shuttering	4	"
8.	Fixing door and window frames	2	"
9.	Rendering internal and external	4	"
10.	Flooring	3	"
11.	Fixing door and window shutters	2	"
12.	Finishing works	5	"

BAR-CHART

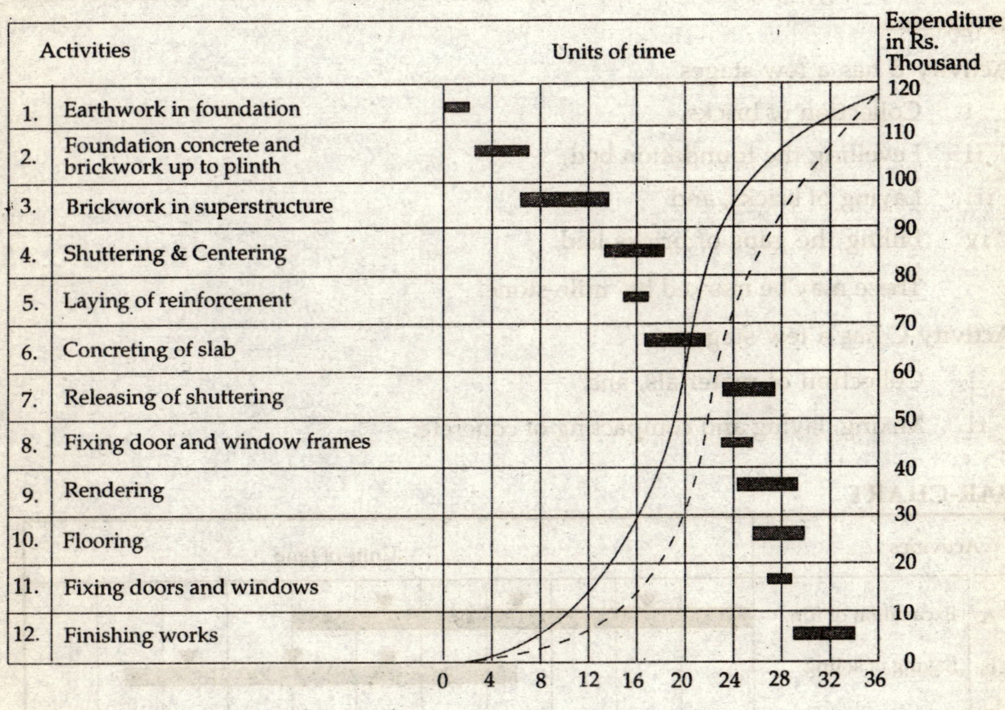

Fig. 2.1 : Bar-Chart

2.3.1 SHORTCOMINGS OF BAR-CHART AND REMEDIAL MEASURES

A Bar-Chart mainly depicts the major items of work and if all the detailed items are shown, the chart would be very clumsy. Moreover, a particular activity — whether small or big—is shown by a bar without any details of sub-activities which would be required for exercising proper control over the project. Further, the actual commencement or completion of an activity with relation to the other is not clear.

These deficiencies can be overcome by introducing marking stages or 'Mile Stones' and by indicating inter-relationship of the activities.

Example.

Let there be a work comprising of the following activities

A.	Excavation of foundation trenches	14	weeks
B.	Laying of soling	12	"
C.	Laying of concrete	8	"

If the activities were taken one after the other, total time required would have been 34 weeks. But the activities can be suitably overlapped and the total duration period can be shortened.

Activity B is dependent upon activity A, again activity C is dependent upon both activities A and B. Activity B can start after activity A has progressed sufficiently. Again, activity C can start when both the activities A and B have progressed.

Activity B has a few stages :

i. Collection of bricks,

ii. Levelling the foundation bed,

iii. Laying of bricks, and

iv. Filling the gaps of bricks laid.

These may be marked by mile-stone.

Activity C has a few stages :

i. Collection of materials, and

ii. Mixing, laying and compacting of concrete.

BAR-CHART

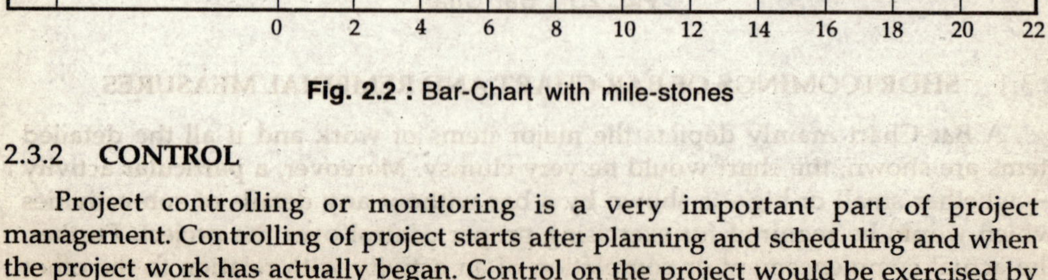

Fig. 2.2 : Bar-Chart with mile-stones

2.3.2 CONTROL

Project controlling or monitoring is a very important part of project management. Controlling of project starts after planning and scheduling and when the project work has actually began. Control on the project would be exercised by following the Bar-Chart diagram as has been drawn earlier. Information collected from the field activity-wise would be imposed on the Bar-Chart or the network diagram periodically and the situation observed if the project is progressing as per programme. If it is observed that the progress is not satisfactory and it is lagging, the cause shall be found out and corrected.

A Bar-Chart, as such, does not show the progress of work, hence cannot be used for controlling the project; but it can be utilised for controlling by showing the programme of work by bars with hatched lines or colouring the bars and the actual progress by introducing another set of bars with different colour and drawn below the programme bars.

Thus a Bar-Chart can be utilised as a tool for controlling a project. Financial programme and progress may be displayed in the Bar-Chart. Control over collection of materials and plants and machineries may thus be effectively exercised by a Bar-Chart.

The manager finds out the lagging activity in the Bar-Chart and takes action for rectification of the cause. In doing so he may have to reschedule the activities, shorten the duration or change the sequence of activity. He may have to reschedule the planning of Resources and Plants and Machineries. The manager needs keeping his watchful eyes on the Bar-Chart. The Bar-Chart will tell the manager the item of work on which his attention is needed. The action required for rectification shall have to be taken as soon as a snag is detected.

2.3.3 ADVANTAGES OF BAR-CHART

i. Bar-Chart is simple, can be drawn quickly and is understood easily. Much training is not required for drawing and understanding it.

ii. Comparison of planned schedule and performance can be had readily by bars drawn showing performances in different colours.

iii. Effects of delay, changes required in cases of emergency can be identified when the activities involved are comparatively less in number.

iv. Interrelationship of activities can be displayed although to a limited extent by introducing 'Mile-stones' or 'marking stages'.

v. Requirement of resources in quantity and exact time when required can be known in advance by separate bars.

vi. Financial position—both planned and performed—can be displayed by introducing S-curve.

vii. Bar-Chart provides for the means of improved communication and co-ordination. Each department is fully aware of the jobs to be done by it and their importance in relation to other jobs. Everyone has the benefit of a clear comprehensive plan which is readily understood. People are more co-operative and are aware of their responsibility much ahead of occurrence.

2.3.4 LIMITATIONS OF BAR-CHART

i. Bar-Chart does not indicate critical activities on whom the attention need be kept, nor does it indicate any warning apprehending any bottleneck in completion of the project.

ii. Bar-chart does not provide for any slack or float period.

iii. Bar-chart itself does not take care of cost aspects of the project unless a S-curve is introduced.

iv. Planning cannot be separated from scheduling in the Bar-Chart.

v. A large number of activities can be started concurrently, but their interdependence cannot be indicated in the Bar-Chart.

vi. The effect of work progressing ahead or lagging of the planned schedule is not indicated.

vii. Reading and reporting on the basis of Bar-Chart is done on the basis of approximate percentage of a particular item represented by a Bar. The information obtained is insufficient for providing any comprehensive idea of the situation.

viii. A Bar-Chart does not indicate the progress of the project as a whole; hence, it cannot be used as a control device which is necessary in cases of a dynamic programme.

ix. Uncertainty or tolerance in duration times and probability are not displayed.

2.4.0 NETWORK MANAGEMENT

The Network Management system uses logical and simple procedures for reducing large volumes of complex information into a graphical or tabular form which can be readily employed in decision making. The techniques enable all round co-ordination among various project agencies to complete the project in time. However, the process need regular review and updating along with the progress of the work.

2.4.1 PERT

PERT — Programme Evaluation and Review Technique. It is basically a method of logical, disciplined and analytical concept of evaluation and reviewing of a programme. It is designed as probablitic network of activities and events. The activities and the network do not take into account the cost involved. However, PERT indicates probable crisis methodology of reporting on the progress.

In this method of programming, the work or the project is divided into a number of clear activities which can be identified separately. Schedule of activities are prepared and represented linearly by short straight lines and are connected with one another by events represented by small circles. The activities are arranged following the sequence of their occurrence and are connected by the events and, thus, a network is formed.

The events either denote starting or ending of an activity. The activities are marked with incoming or outgoing arrow signs from the nodes to indicate their progress direction. Each activity requires some period of duration for its completion. This time period is noted above the activity line at the middle.

The expected time period of duration of an activity is estimated by evaluation of mean of the three times as below :

$$t_e = \frac{1}{6}(a + 4m + b)$$

where, t_e = expected time.

a = optimistic time, the time that would be required for completion of the activity if nothing goes wrong.

b = pessimistic time, the time that would be required for completion of the activity taking into account all sorts of difficulties and bottlenecks.

m = most likely time, time for completion of the activity which occurs most frequently.

Example

Where, a = 3, b = 15 and m = 6

$$t_e = \frac{1}{6}(3 + 4 \times 6 + 15)$$

$$= \frac{1}{6}(3 + 24 + 15)$$

$$= \frac{42}{6}$$

$$= 7$$

The activities of example in paragraph 2.3 is reproduced in network.

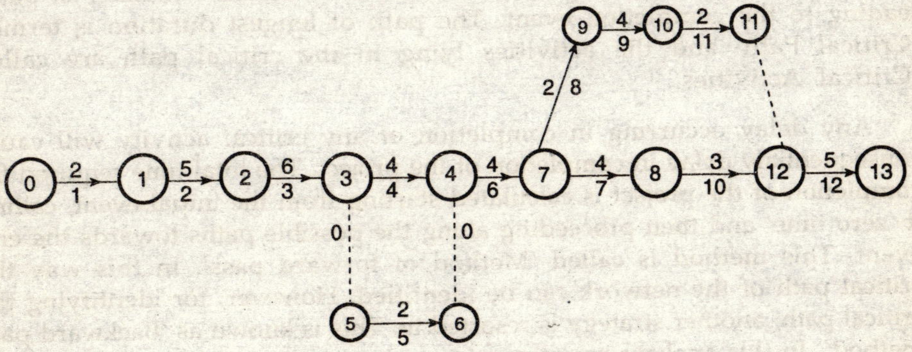

Fig. 2.3 : Network Diagram

2.4.2 ADVANTAGES OF PERT

PERT provides for a number of significant advantages. Some of them are mentioned below :

i. The most important advantage of the technique is the possible anticipation of crisis. Warning can be given sufficient time in advance and enough time is generally available for correction of the process of the work.

ii. PERT also signals need to the completion dates of some activity so that they can be matched with schedule of other activity. Thus it may help in retardation of progress in the department over producing and causing blockage of capital. It can guide the management in deployment of labour according to necessity.

iii. PERT provides for logical and disciplined basis for planning. Breaking the project into activities and finding out their relationship to build a network is much helpful as it clarifies the thinking of the management as to what is involved in the project. The project as a whole could be visualised clearly at any time. This induces possible improvement with the resultant saving of time.

iv. PERT provides for precise reporting of the progress. Delays can be pin-pointed and responsibilities fixed up. Better understanding of the project status and related various departments is possible.

2.5.0. CRITICAL PATH METHOD, CPM

CPM means Critical Path Method. It is synthesising in concept. The network diagram is activity oriented and deterministic in nature. In CPM network, time is related to cost.

Critical Path

A network of activities are drawn for programming of the project for finding out the possible completion time of any intermediate event and to estimate the total project time.

In estimating the total project time, there would be a number of paths leading to the completion event. The path of longest duration is termed 'Critical Path' and the activities lying in the critical path are called 'Critical Activities'.

Any delay occurring in completion of any critical activity will cause consequentialy delay in completion of the project. The total time required for completion of the project is calculated starting from the initial event, calling it 'zero time' and then proceeding along the possible paths towards the end event. This method is called 'Method of forward pass'. In this way the critical path of the network can be identified. However, for identifying the critical path, another strategy is resorted to. This is known as 'Backward pass method'. In this analysis counting of time begins from the last event whose occurence of earliest time (E) is known. The latest possible occuraence times (L) of the events are calculated by subtracting the times. The strategy of backward pass may be extended to the entire network. When the exercise is complete, it will be seen that in a critical path E and L values of the tail and head events of all the critical activities are the same and the difference between the E values of the tail and head events and the L values of the head and tail events are the same and equal to the duration period of the activities.

The activities of example in paragraph 2.3 is reproduced in network.

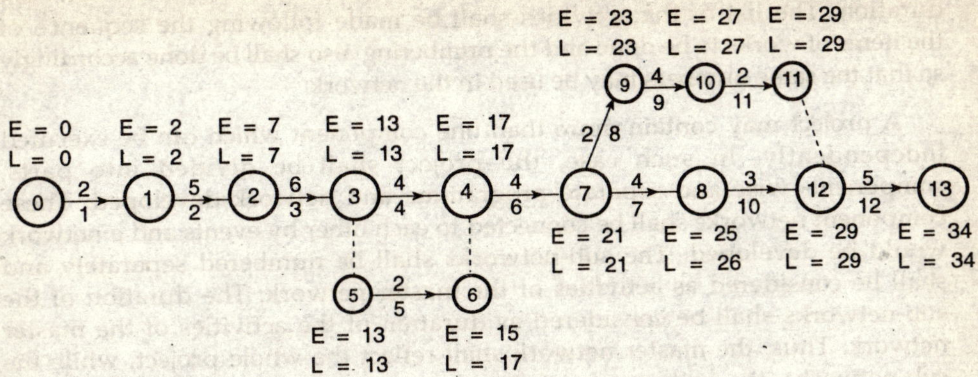

Fig. 2.4 : Network Diagram with Critical Path

Path I Activity	Time	Path II Activity	Time	Path III Activity	Time	Path IV Activity	Time	weeks
0 – 1	2	0 – 1	2	0 – 1	2	0 – 1	2	"
1 – 2	5	1 – 2	5	1 – 2	5	1 – 2	5	"
2 – 3	6	2 – 3	6	2 – 3	6	2 – 3	6	"
3 – 4	4	5 – 6	2	5 – 6	2	3 – 4	4	"
4 – 7	4	4 – 7	4	4 – 7	4	4 – 7	4	"
7 – 8	4	7 – 9	2	7 – 8	4	7 – 9	2	"
8 – 12	3	9 – 10	4	8 – 12	3	9 – 10	4	"
12 – 13	5	10 – 11	2	12 – 13	5	10 – 11	2	"
		12 – 13	5			12 – 13	5	"
	33		32		31		34	weeks

PATH IV is critical.

Programme prepared by a network shall be strictly adhered to for completion of the project in the prescribed time period.

However, in practice, it would be seen that due to various reasons there may be delay in completion of a particular activity. In such case, changes would occur in the network and the critical path may be altered and some other activities may become critical. These would be dealt in the control of the project.

2.5.1 SCHEDULING

i. Scheduling of the activities.

Scheduling of activities involved are required for drawing of the programme; hence it should be drawn prior to programming. Scheduling of activities comprises of list of the activities with activity number and their duration. The list of the activities shall be made following the sequence of the items of works to be done and the numbering also shall be done accordingly so that the same numbers may be used in the network.

A project may contain more than one component which can be executed independently. In such case, the project shall be divided into parts, component- wise and separate programme and network developed. These component networks shall be connected to each other by events and a network would be developed. The sub-networks shall be numbered separately and shall be considered as activities of the master network. The duration of the sub-networks shall be considered as duration of the activities of the master network. Thus, the master network shall reflect the whole project, while the sub-networks, the programme of the component units. (Also see paragraph 3.2.1).

ii. Scheduling of Resources (Materials)

When a programme of the project has been drawn either by Bar-Chart or by Network diagram, the period and time of execution of a particular activity or item of work is known. The estimate of the project with all details has been prepared earlier, the volume of work of the activity and the resources or materials that would be required is known. From these information, schedule or schedules of resources with their quantities and the time when they would be required can be prepared.

This can again be represented in a Bar-Chart or a network and can always be compared with the main Bar-Chart or network diagram of the work proper.

iii. Scheduling of Plants and Machineries

Various plants and machineries may be required for a project at different stages of the progress. For example, concrete mixing machine, vibrator would be required for concreting work.

The volume of work and period of the items of works are known from the programme. The particular plant and machineries that would be required can be found out and the period for which these would be required can also be worked out. A Bar-Chart or a network diagram can be prepared for the plants and machineries so that they are available at site at the time of necessity. Care shall be taken to incorporate spare number of plants and machineries so that there may not be any hold – up of work during actual execution due to non-availability of the plants and machinery.

Chapter 3

Line of Balance

3.1.0 LINE OF BALANCE

Line of balance is basically a method for scheduling and displaying the progress of repetitive manufacturing programme in industries and shops.

In a continuous manufacturing process, a product moves down a line, passes successive work stations where one after another operation is done and it takes final form. This system is called line of production. Line of balance method is based on this principle. In most of the assembly shops, the work progresses in this system. Conveyor belt is sometimes used for further improvement and sophistication and for convenience, speed and avoiding defect.

In a line production system the following are important necessities :

i. Identifiable milestones in production line for exercising better control,

ii. Predetermined and uninterrupted duration time of each activity,

iii. Uninterrupted resource supply, and

iv. Predetermined delivery schedule.

Application

The line of balance was primarily developed for monitoring production jobs in factories having repetitive nature of items. The method can suitably be introduced in construction management having repetitive items of work. For example, mass housing construction of type buildings, where same types of houses are built in large numbers. The method is effectively employed in production of prefabricated housing elements in housing factories.

Technique

The technique is based on four elements as below :

i. **Objective chart**

Objective of the project with preplanned delivery schedule and targets;

ii. **Programme chart**

A chart indicating the sequences of activities and duration time;

iii. **Progress chart**

A progress chart showing the physical achievement of the activities for exercising control at various points;

iv. **Line of Balance**

The Line of Balance is drawn across the progress chart shown by vertical bars, co-ordinating the programme plan and the objective. The Line of Balance indicates the predetermined levels of activities at each control point. The progress having been indicated against LOB (Line of Balance), will indicate the position of performance in relation to programme at each control point.

However, LOB predicts only the physical status of production in relation to the schedule. Assessment of whether the progress is ahead or lagging need be made and necessary input need be sent to the concerned department for correction.

3.1.1 PROCEDURE OF DRAWING LINE OF BALANCE

i. **Objective chart**

A delivery schedule of the items together with a cumulative schedule is prepared and then it should be drawn graphically with relation to delivery time.

ii. **Programme chart**

A chart showing all the parts or activities with number of pieces required per unit for completion with duration of time for production is prepared;

With the help of the chart a network is drawn of the activities according to sequence showing the duration time below the activity line and number of parts required per complete item in small circles. The events are then numbered starting from the end event.

iii. **Progress chart**

A chart showing physical progress of the activities is to be prepared of all the activities on the particular point of time.

iv. **Line of Balance**

As there may be fluctuating delivery schedule and time in manufacture, production level of each activity need be determined. The number of units of production will vary for each activity. A chart showing the calculation of balance of quantities of each activity at the end of the point of time is to be developed. This will depict activity with number, quantity required for each product, lead of time, number of products and total number of products.

Vertical bars corresponding to relative time lag of that activity is to be drawn and erected against each activity at which the bar meets the curve will indicate the planned cumulative level of that activity. It is recorded by drawing a horizontal line.

The horizontal lines thus obtained on each activity when added together gives the 'Line of Balance'.

Control and action

The number of units above the LOB shows over production and anything below the line shows under production at that point of time. The chart enables the management to take necessary action for correcting measures needed for the particular production line. Thus LOB gives an early warning to the management and can be used for functional purpose also. The LOB chart can be tabulated and proposed action for correcting the balance can be noted against each activity.

Advantages of programming by LOB method

a. The LOB method enables the management to report on all controlling points in the programme and to identify the lead time and time coverage for each.

b. It gives visual idea of the position of a project as a whole at any point and facilitates remedial measures to be taken.

c. It helps in recasting of the programme, if certain planning do not come true. However, LOB predicts the condition of attaining or failing of the schedule. There is no assessment of how far ahead or behind the total project might be. There is limited adaptability or indication where and how much improvement can be made to rectify the schedule slippage.

Example

A carpentry shop is to produce 2,000 pieces of single leaf panelled shutters having two panels. The shutters in complete state with a coat of priming is to be delivered in 20 weeks as per delivery schedule. Draw LOB on the 5th week of the delivery schedule.

Objective Chart – Delivery Schedule

Week number	Delivery per week	Cumulative quantity
1	50	50
2	50	100
3	100	200
4	100	300
5	100	400
6	100	500
7	100	600
8	150	750
9	150	900
10	150	1,050
11	150	1,200
12	100	1,300
13	100	1,400
14	100	1,500

Week number	Delivery per week	Cumulative quantity
15	100	1,600
16	100	1,700
17	100	1,800
18	100	1,900
19	50	1,950
20	50	2,000

DELIVERY CHART

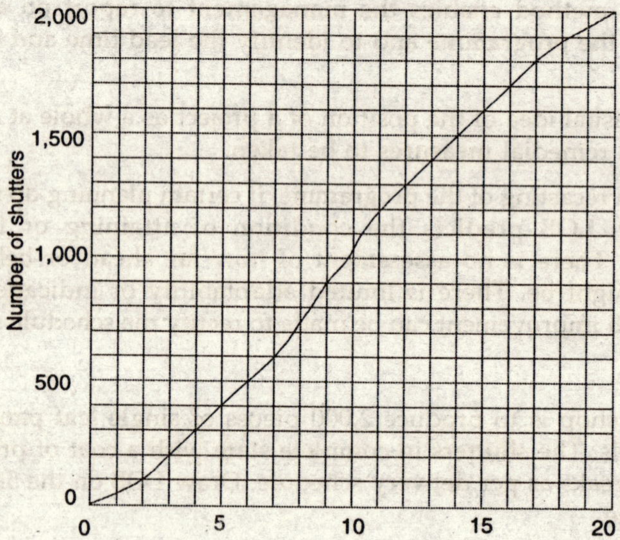

Fig. 3.1 : Delivery chart—line of balance

The next step is to prepare a programme chart which comprises of two parts. List of activities with number required for each door shutter and duration as per network.

Programme Chart

Activity number	Name of the Activity	Number required for each leaf	Duration
1	Top rail	1	2 weeks
2	Lock rail	1	2 "
3	Bottom rail	1	2 "
4	Vertical style	2	4 "
5	Top panel	1	2 "
6	Bottom panel	1	2 "
7	Assembly of rails and styles	1	4 "

Activity number	Name of the Activity	Number required for each leaf	Duration
8	Assembly of panels	1	4 weeks
9	Finishing of shutters	1	4 "
10	Painting one coat	1	2 "

Net work – Calculation of lead times

⑤ Top rail (1) / 2 ③

⑤ Lock rail (1) / 2 ③ Assy. of rails & styles (1) / 4 ② Finish (1) / 4 ① Painting (1) / 2 ⓪

⑤ Bottom rail (1) / 2 ③

⑤ Vert. Styles (2) / 4 ③

⑥ Top panel (1) / 2 ④ Assy. of panels (1) / 4 ② Note : ( ) indicates number required per shutter.

⑦ Bottom panel (1) / 2 ④

Fig. 3.2 : Network – Calculation of lead time.

Progress chart —Performance

Week number	Cumulative delivery programme	Actual delivery in the week	Cumulative delivery
1	50	50	50
2	100	50	100
3	200	100	200
4	300	100	300
5	400	100	400
6	500	100	
7	600		
8	750		
9	900		
10	1,050		
11	1,200		
12	1,300		
13	1,400		
14	1,500		
15	1,600		

16	1,700
17	1,800
18	1,900
19	1,950
20	2,000

Calculation of Balance quantities

Activity No.	Qty. per shutter	Total lead time (week)	Number of shutters	Total Qty. of parts	Performance at week 5
1. Top Rail	1	12	1,800	1,800	2,000
2. Lock rail	1	12	1,800	1,800	2,000
3. Bottom rail	1	12	1,800	1,800	1,800
4. Vertical styles	2	14	1,950	3,900	3,900
5. Top panel	1	12	1,800	1,800	1,600
6. Bottom panel	1	12	1,800	1,800	1,500
7. Assembly of rails and styles	1	10	1,600	1,600	1,600
8. Assembly of panels	1	10	1,600	1,600	1,500
9. Finishing	1	6	1,200	1,200	1,200
10. Painting	1	2	600	600	600

Calculation of balance quantities at the week 5

Supposing that week 5 of the delivery schedule has been reached and we are to find the balance of quantities and draw the LOB. Let us take the part no. 1—Top rail, the lead of time is 12 weeks and the total lead is 12 + 5 = 17 weeks. In 17 weeks, the production level of the complete shutters would be 1,800 numbers as per chart and graph. As only 1 number part no. 1—Top rail would be required per shutter, the production level of the Top rail would be 1,800 × 1 = 1,800 numbers. Similarly, as 2 numbers of vertical styles (part no. 4) are required per shutter, production levels of part no. 4 would be 3,900. In this way, the production level of all the activities are found out and plotted by drawing horizontal lines across the respective vertical bars of activities. The horizontal lines when joined would give the LOB.

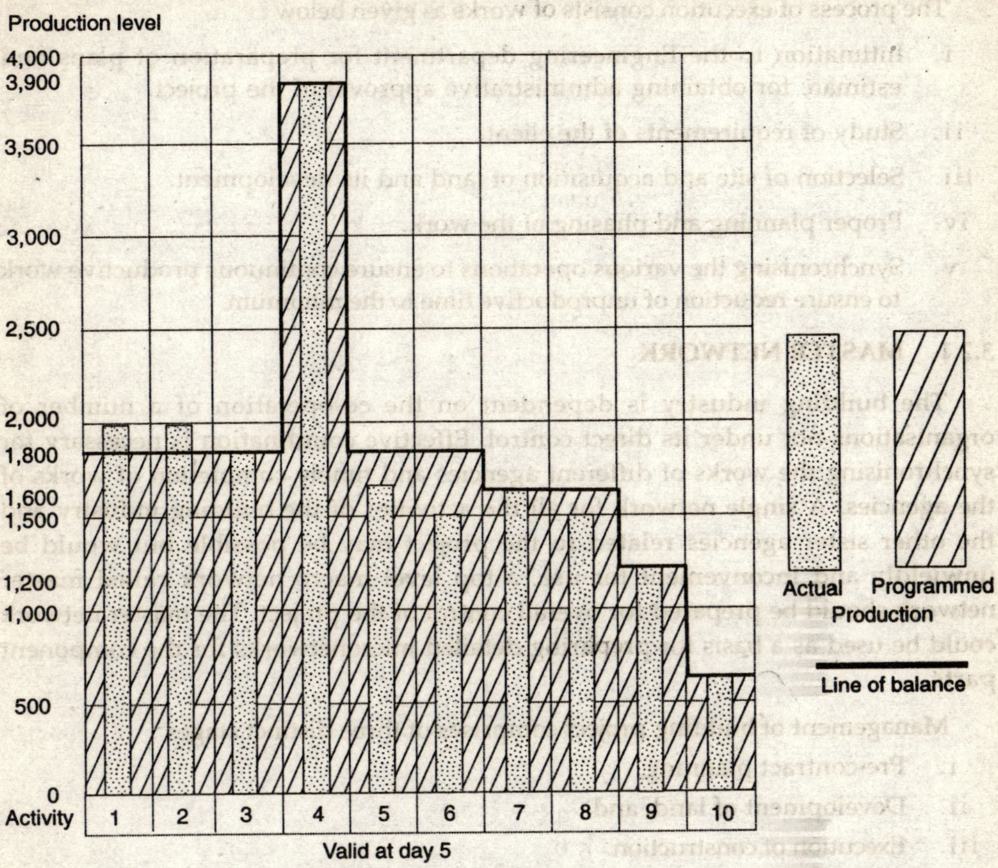

Fig. 3.3 : Line of Balance Diagram.

Observations :			**Actions**
Activity nos.	1 and 2 – Over-production,		Action required for retarding.
Activity nos.	3, 4, 7, 9 & 10	at par	—
Activity nos.	5, 6 and 8	lagging.	Action required for expediting.

3.2.0 INTEGRATED MANAGEMENT SYSTEM FOR BUILDING PROJECT

Application of management techniques like PERT, CPM is needed for achieving the primary aim of management – utilisation of financial, human, machine and material resources effectively, fruitfully and thus obtaining the end product economically.

CBRI has applied the Network Technique in different building projects and has worked out an Integrated Management System (IMS), suitable in Indian conditions. The system integrates all the different activities in the process of execution of a building project – Planning and Control from the inception to completion.

The process of execution consists of works as given below :

i. Intimation to the Engineering department for preparation of plans and estimate for obtaining administrative approval of the project.

ii. Study of requirements of the client.

iii. Selection of site and acquisition of land and its development.

iv. Proper planning and phasing of the work.

v. Synchronising the various operations to ensure continuous productive work to ensure reduction of unproductive time to the minimum.

3.2.1 MASTER NETWORK

The building industry is dependent on the co-operation of a number of organisations not under its direct control. Effective coordination is necessary for synchronising the works of different agencies and timely completion of works of the agencies. A single network for all the activities of the building industry and the other sister agencies related to the project may be possible but would be unwieldly and inconvenient for use, a top level macro network called master network should be prepared for overall control of the project. The master network could be used as a basis for preparing detailed micronetworks for the component parts.

Management of building project comprises of three distinct stages :

i. Pre-contract planning,

ii. Development of land, and

iii. Execution of construction.

For all these stages separate network should be prepared.

3.2.2 FLOW OF INFORMATION

In order to avoid delays due to non-availability of drawings, a check list for flow of information about (a) the various drawings needed at the design and estimate stage and (b) working drawings required during execution should be prepared on a time schedule.

3.2.3 MATERIAL MANAGEMENT

To overcome shortage and delays due to materials and labour, management chart should be prepared giving details of movement of (i) materials and (ii) man at site.

3.2.4 PLANNING OF MASS HOUSING

Planning of Mass Housing Project comprising of a large number of buildings of the same type is done by integrating a Critical Path Network of one building, establishing the critical activities and then Line of Balance Technology is applied for the repetitive blocks. Since, due to several unforseen reasons, the work cannot strictly proceed as planned, it is advisable to allow for some reserve/rest time between each activity.

An illustrative example of a typical line of balance technology planning for mass housing keeping continuity of labour gangs for repetitive work is shown in figure 3.4.

3.2.5. HOW TO DRAW LINE OF BALANCE TECHNOLOGY PLANNING FOR MASS HOUSING

Example as shown in Figure 3.4 Ten buildings of same type is taken into consideration.

i. A Critical Path Network of a type building is drawn separately and its critical activities are identified.

ii. Straight line logic of critical activities is drawn. The activities are all critical. The activities comprise of all related activities needed for completing the critical activities.

iii. Labour deployment needed for completion of the critical activities is drawn in straight line showing the critical activities.

iv. Straight line critical path of each building needing 244 days for completion are drawn for ten buildings. The first straight line path starts from 0 time and ends at 244 days. The second one starts after an interval of 12 days and so on. The straight line is taken as base and is divided into equal divisions of 10 days each.

v. The vertical line at 0 is divided into 9 equal divisions. At each dividing point, ten horizontal lines are drawn, each representing a critical path of building.

vi. A delivery schedule of the buildings is drawn.

Delivery schedule of the buildings

Building number	Starting date	Delivery date (days)
1	0	244
2	12	256
3	24	268
4	36	280
5	48	292
6	60	304
7	72	316
8	84	328
9	96	340
10	108	352

vii. For finishing works 36 days have been allotted. The working days have been shown from 135 days to 244 days for the building number 1. During this period the finishing works will be done, but the total time required would be 36 days.

viii. Gang formation table is formed.

Gang No.	Gang Index	Activities	Duration (days)
1	g_1	Foundation earthwork	12
2	g_2	Foundation brickwork	12
3	g_3	Brick work 1st floor	24
4	g_4	Brick work 1st floor.	24
5	g_5	First floor slab	12
6	g_6	Second floor brickwork	24
7	g_7	Second floor brickwork	24
8	g_8	Second floor slab	12
9	g_9	Third floor brickwork	24
10	g_{10}		
11	g_{11}	Third floor slab	12
	$g_{6,7}$	Finishing works	36

ix. Deployment of gangs of labours are made in multiple of an unit time suitably fixed for the gangs e.g. completion of foundation works is estimated to need 24 days. This is to be done by two gangs. Gang no. 1 would do the excavation of the foundation trench, Gang no. 2 would do the concreting, brickwork and D.P.C. Thus the foundation work of building no. 1 will be complete in 12 + 4 + 12 = 28 days, with 4 days for accommodation of rest time or any slip. When gang no. 1 has completed their job of foundation of building no. 1, they would shift to building no. 2 just after 12 days. So, work of each building would start at regular interval of 12 days. Building no. 1 would be completed in 244 days with provision of 4 × 10 = 40 days for Rest time or slip time. Thus delivery of the buildings will be at 12 days' interval. The last building will be delivered after 352 days.

Shifting of the gangs has been shown in dotted lines.

Thus the whole project would be completed in 352 days.

In addition to the activities as described above, short term programmes for 3 to 4 months, cash inflow and outflow charts and weekly programmes shall be prepared.

Even those who are not familiar with drawing of the network and critical path analysis would be able to use the Integrated Management System to achieve efficiency and economy.

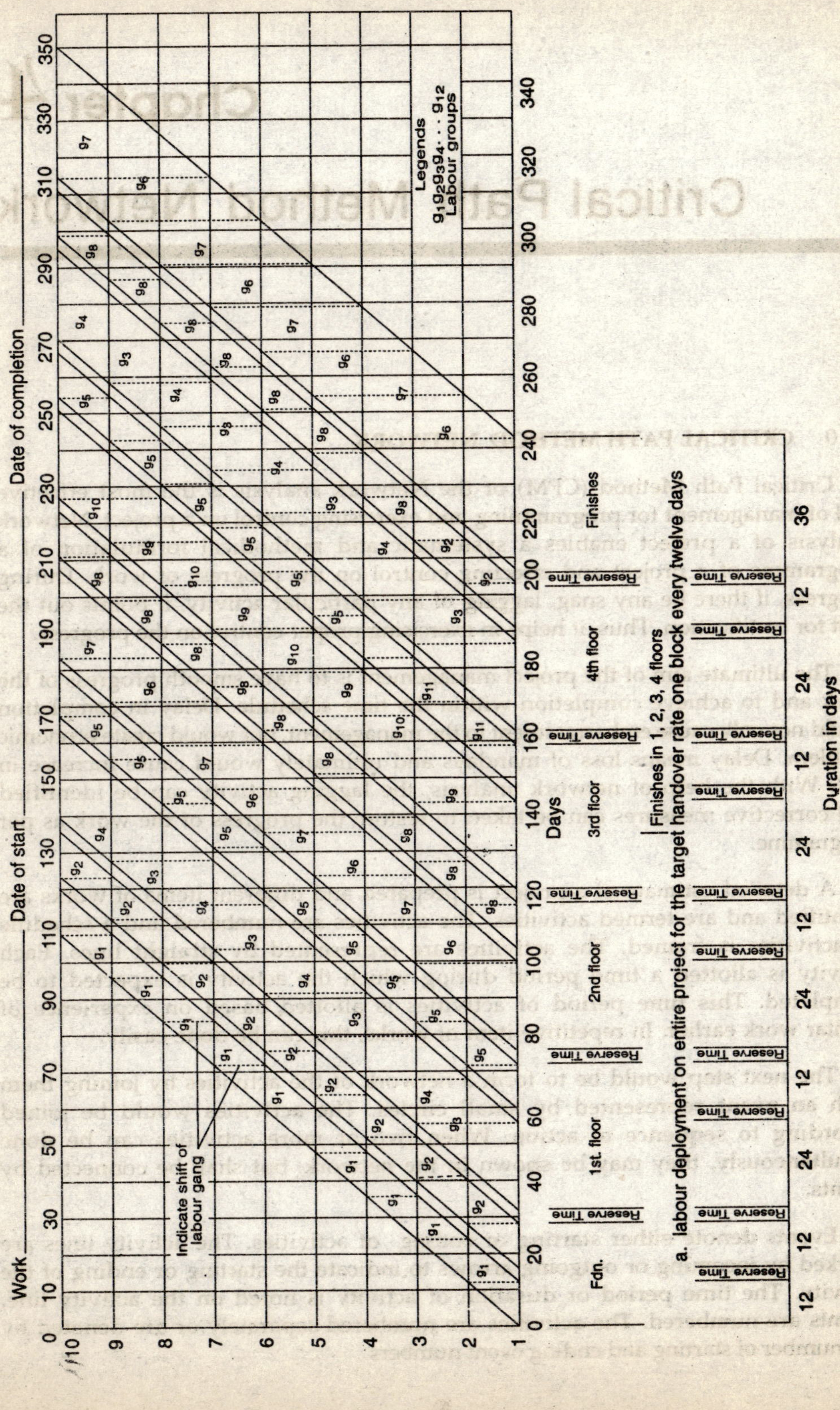

Fig. 3.4 : Line of balance — Planning for mass housing.

Chapter 4

Critical Path Method Network

4.1.0 CRITICAL PATH METHOD NETWORK

Critical Path Method (CPM) or the Network analysis is the most effective tool of management for programming and exercising control on a project. Network analysis of a project enables a systematic and methodical formulation of a programme of a project and effecting control on the progress of work. During progress, if there be any snag, lagging of any particular activity, it points out the spot for rectification. Thus, it helps in exercising proper control on the progress.

The ultimate aim of the project management is to have smooth progress of the work and to achieve completion within the time schedule. Delay in completion would not only cause embarrassment to the management, but would create economic problem. Delay means loss of mandays and ultimately would cause increase in cost. With the help of network analysis, the lagging activity can be identified and corrective measures can be taken to restore the progress of the work as per programme.

A detailed estimate of a project is prepared and different items of works are identified and are termed activities. The activities are numbered and a schedule of activities is framed. The activities are represented by straight lines. Each activity is allotted a time period during which the activity is expected to be completed. This time period of activities is allotted, based on experience of similar work earlier. In repetitive item of works, this can be done easily.

The next step would be to form a network of the activities by joining them with an event represented by small circles. The activities would be joined according to sequence of action. When two or more activities can be done simultaneously, they may be shown in the network, but shall be connected by events.

Events denote either starting or ending of activities. The activity lines are marked by incoming or outgoing arrows to indicate the starting or ending of the activity. The time period or duration of activity is noted on the activity line. Events are numbered. The activities are numbered separately or are denoted by the number of starting and ending event numbers.

Example

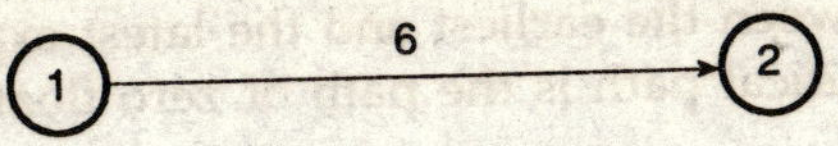

Fig. 4.1 : Activity

Here, the activity 1 – 2 with duration of 6 units of time is shown.

4.1.1 DUMMY

When activities can proceed independently, but their starting or completion is dependent on starting or completion of some other activity, the actions can be shown by a dummy activity which has 0 time.

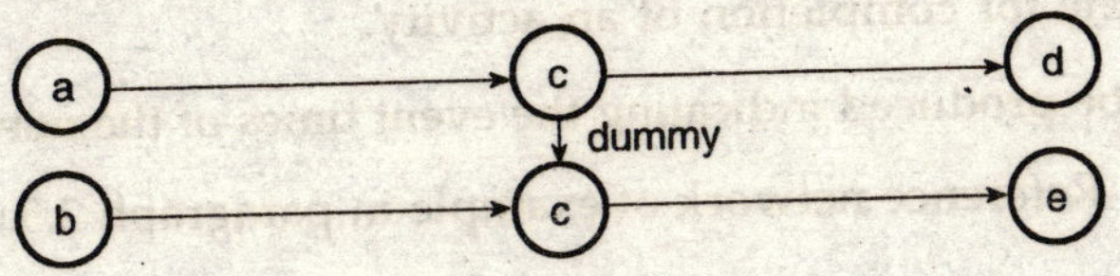

Fig. 4.2 : Dummy

Here, activity a– c and b – c can begin independently, but the activities c – d and c – e can begin only when activities a – c and b – c are complete. To represent the situation, dummy activity c – c with 0 time period has been introduced.

The arrow-heads indicate the completion of activities. The beginning of an activity is called a 'tail event' and the completion of an activity is called a 'head event' i.e. activity moves from 'tail event' to 'head event' or 'completion event'.

Time flows from left to right, and 'head events' always have a number higher than that of the 'tail event'.

4.2.0 OCCURRENCE OF ACTIVITIES

The list of activities and the network diagram of example in paragraph 2.5.0 may be referred to.

It will be seen that there are four possible paths for reaching at the completion event number 13. Each path requiring different total time. The longest path is number IV, requiring total duration of 34 weeks. The longest path in a network is called critical path and the activities lying in the critical path are called critical activities.

The occurring time of the events had been calculated by forward pass i.e. the times of occurrence of the events have been calculated from the starting event to the end event proceeding to the forward direction. These times are the earliest occurring times. The earliest and the latest occurring time of the end event would be same. Now, occurring time of the events may be calculated by backward pass, starting from the end event. The times now available are the latest occurring times. The earliest occurring times and the latest occurring times of the events are noted in the network. It will be seen that the earliest and the latest occurring times of the events in the critical path are the same.

4.3.0 FLOAT

The difference between the earliest and the latest expected times of an event is termed float. The critical path is the path of zero float.

i. An earliest starting time which is the earliest possible time when an activity can start.

ii. The earliest finish time which is the earliest possible time when an activity can be finished.

iii. The latest starting time which is the latest possible time when an activity can start.

iv. The latest finish time which is the event time of the head event and the latest time for completion of an activity.

A table can be produced indicating the event times of the network.

Reference network of example in paragraph 2.5.0

Activity	Duration	Starting time		Finishing time	
		E	L	E	L
0 – 1	2	0	0	2	2
1 – 2	5	2	2	7	7
2 – 3	6	7	7	13	13
3 – 4	4	13	13	17	17
5 – 6	2	13	13	15	17
4 – 7	4	17	17	21	21
7 – 8	4	21	21	25	26
7 – 9	2	21	21	23	23
9 – 10	4	23	23	27	27
10 – 11	2	27	27	29	29
8 – 12	3	25	26	29	29
12 – 13	5	29	29	34	34

4.3.1 TOTAL FLOAT

Let us consider the activities numbering 5 – 6, 7 – 8 and 8 – 12 of the table.

Activity number 5 – 6

Maximum available time = 17– 13 = 4 weeks.

Duration = 2 weeks

Total float = available time – duration

= 4 – 2 = 2 weeks

In the same way,

Activity number 7 – 8

Maximum available time	= 26 – 21	= 5 weeks
Duration	= 4 weeks	
Total float	= 5 – 4	= 1 week
Activity number	= 8 – 12	
Maximum available time	= 29 – 25	= 4 weeks
Duration		= 3 weeks

Total float = 4 – 3 = 1 week.

For all other activities the maximum available time and duration are the same, hence no total float is available.

4.3.2 FREE FLOAT

Float in activity would be reduced if event 5 cannot start unless after 4 weeks i.e. the maximum available time.

Hence, the free float is given by

Free float (FF) = Earliest event time of event 6 – Earliest event time of 5 – duration.

= 15 – 13 – 2 = 0

4.3.3 INDEPENDENT FLOAT

In some cases, the absorption of float affects neither the predecessing nor the successing activities. The float is then termed 'Independent float'.

Independent float (IF) = Earliest event time of 6 – latest event time of 5 – duration

= Minimum available time – duration

= 15 – 13 – 2

= 0

Hence, the floats are

	Total	Free	Independent
Activity 5 – 6	2 weeks	0	0

4.3.4 SIGNIFICANCE OF FLOAT

i. The critical path in a network is that path which has zero float.

ii. Float presents under-utilised resources, and

iii. Float represents flexibility of an activity and disappearance of float means less flexibility for the activity.

4.4.0 PROJECT COST BY CPM

Total project cost consists of

i. Indirect costs, and

ii. Direct costs.

Again, indirect project costs are of two types

i. Fixed indirect costs, and

ii. Variable indirect costs.

Fixed indirect costs include costs which must be met independentof the progress of the work. These are statutory costs. Variable indirect costs include all costs which vary more or less directly with the volume of work performed. This varies with time. These include overhead expenditure, interest on capital, amenities etc.

Analysis for direct cost is more complex. Direct cost is the amount of cost which is directly dependent on the amount of resources involved during a period of time. The resources include labour, materials, plant and machineries, payment to sub-contractor etc.

Generally, it is seen that for an activity, indirect cost increases with time while direct cost decreases. Hence, it is probable that a minimum may exist for total cost for certain value of activity time.

Sometimes, it becomes necessary to crash duration time of an activity for achieving certain goal. Due to crashing of duration time of an activity cost increases. The graphical relation would explain the situation.

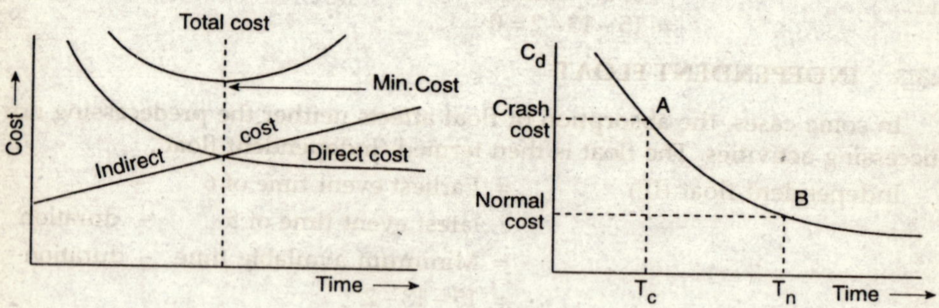

Fig. 4.3 : Graphical relation of Cost & Time

4.5.0 CRASHING A NETWORK FOR OPTIMUM PROJECT COST

In a multiple activity network, the total project cost is calculated on activity to activity basis. Usually, the gain in indirect cost is offset by increase in direct cost due to reduction in project time.

Procecure followed for crashing

i. When expediting of the project is essential, i.e. total project time desired is less than estimated project time based on normal conditions, crashing of activity time may be resorted to.

ii. Project time is determined by adding all the activity times lying in the critical path, some of these critical activities have to be 'crashed' for expenditing the project.

iii. Crashing is carried out with one activity at a time unless any other linked activity is affected due to crashing of the one.

iv. The activity to be crashed is decided from the index,

$$I = m - \frac{C_{dc} - C_{dn}}{T_n - T_c} \quad I \gtrless 0, \text{ where } m = \text{marginal cost}$$

is maximum.

v. Crashing begins from 'all normal' condition and proceeds step by step to 'all crash' solution.

vi. During this step by step process of crashing, it may be found that for some overall project time, the project cost is minimum. This project time is the optimised project time for optimum cost.

vii. Keeping the optimum time same, the cost can further be reduced by decompression of the non-critical activities, if possible without changing the critical path.

Direct cost of expediting an activity per unit time is given by

$$\frac{\text{Crash cost} - \text{Normal cost}}{\text{Normal time} - \text{Crash time}}$$

Direct cost at any discrete time, T is given by,

$$C_d = \frac{C_{dc} - C_{dn}}{T_n - T_c} T$$

4.5.1 CRASHING A PROGRAMME

Let us consider the network given below :

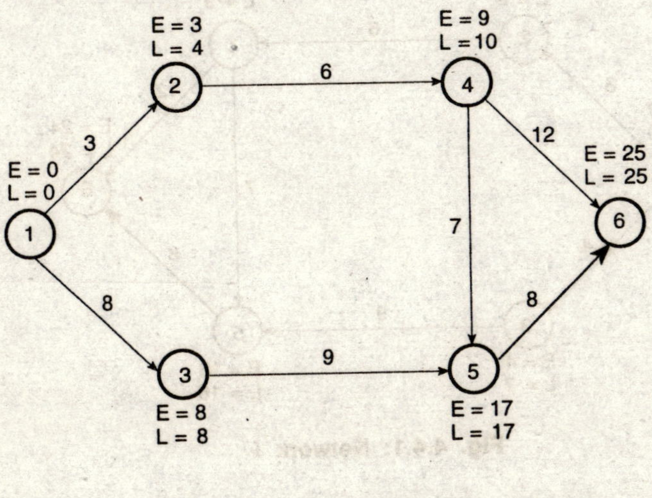

Fig. 4.4 : Network Diagram—Normal

The network for normal duration times indicate the project time of 25 days with critical path 1—3—5—6.

The utility data for the network are tabulated as below.

Activity	Normal duration time	Crash duration time	Normal cost	Crash cost	$\frac{C_{dc} - C_{dn}}{T_n - T_c}$
	(days)	(days)	(Rs.)	(Rs.)	(Rs. per day)
1–2	3	2	1,000	1,070	70
1–3	8	4	500	700	50
2–4	6	2	700	1,000	75
3–5	9	3	1,200	1,800	100
4–5	7	2	600	1,000	80
4–6	12	4	700	1,180	60
5–6	8	3	800	1,150	70

Let the indirect cost per day be Rs. 80/-

The all normal project cost = Normal cost + duration × indirect cost per day

= Rs. 5,500/= + 25 × 80

= Rs. 7,500/=

The maximum value of I in the critical path is for activity 1 – 3.

I. Crash activity 1 – 3

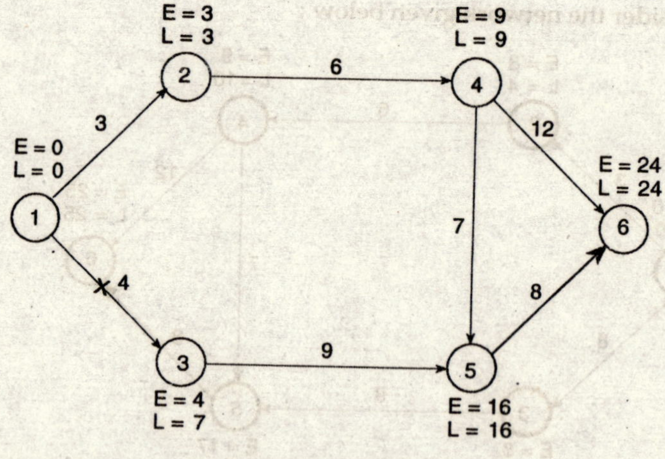

Fig. 4.4.1: Network I

The critical path is changed to 1—2—4—5—6

Duration is 24 days.

New project cost = Rs. 7,500/= + 4 × 50 – 1 × 80

= Rs. 7,620/=

II. Crash activity 1 – 2

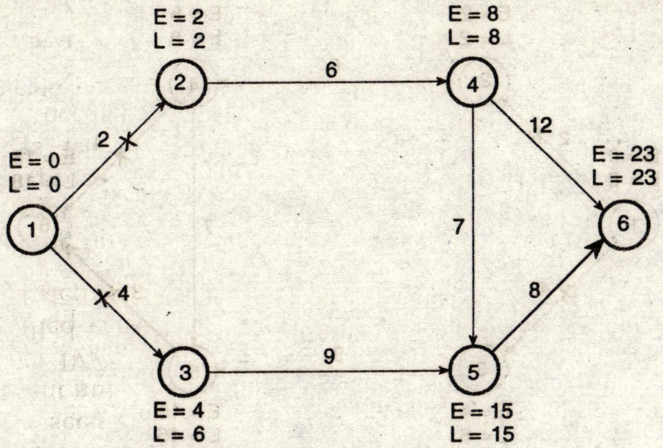

Fig. 4.4.2 : Network II

The critical path remains unchanged 1—2—4—5—6

Duration is 23 days

New project cost = Rs. 7,620/= + 1 × 70 – 1 × 80

= Rs. 7,610/=

III. Crash activity 5 – 6

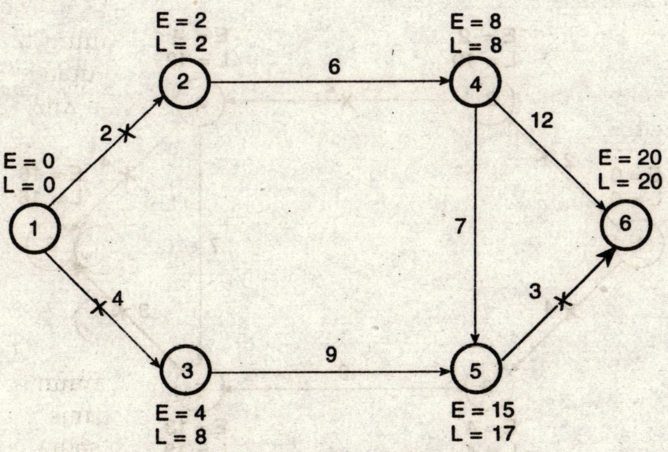

Fig. 4.4.3 : Network III

Critical path is changed to 1—2—4—6

Duration is 20 days

New project cost = Rs. 7,610 + 5 × 70 – 3 × 80

= Rs. 7,720/=

IV. Crash activity 4 – 6

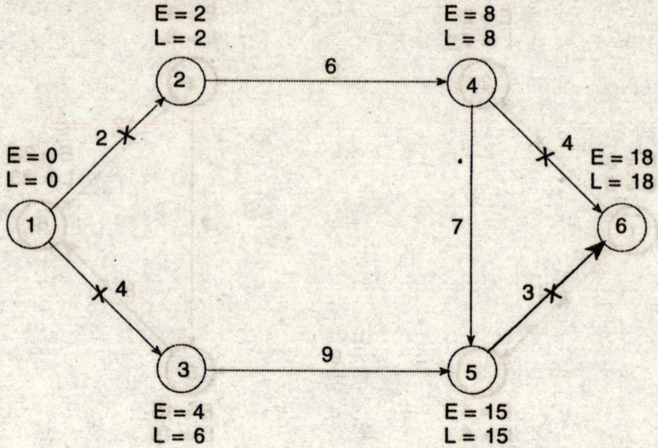

Fig. 4.4.4 : Network IV

Critical path is changed to 1—2—4—5—6

Duration is 18 days

New project cost = Rs. 7,720/= + 8 × 60 – 2 × 80

= Rs. 8,040/=

V. Crash activity 2– 4

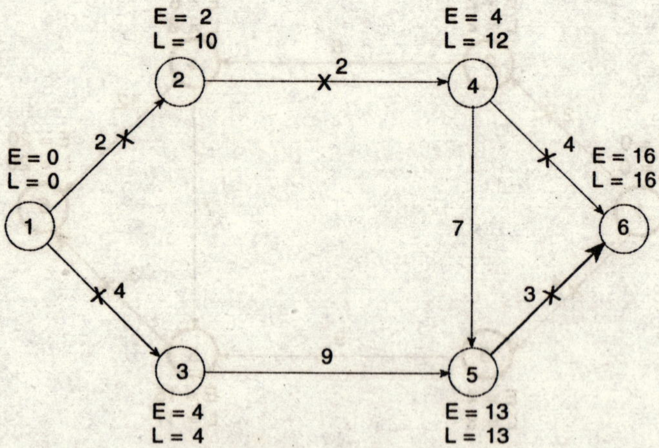

Fig. 4.4.5 : Network V

Critical path is changed to 1—3—5—6

Duration is 16 days

New project cost = Rs. 8,040/= + 4 × 75 – 2 × 80

= Rs. 8,180/=

VI. Crash activity 3 – 5

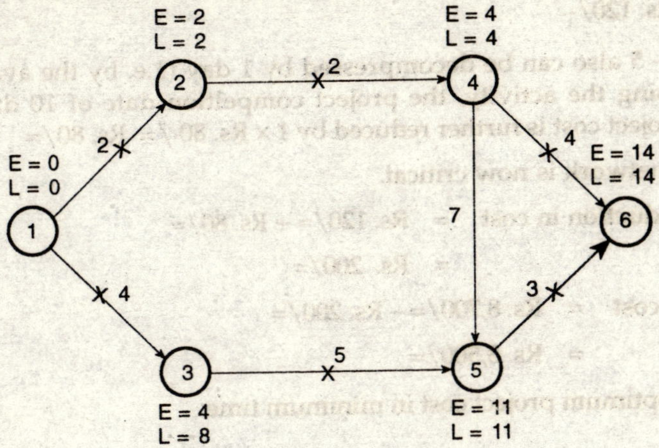

Fig. 4.4.6 : Network VI

Critical path is changed to 1—2—4—5—6

Duration is 14 days

New project cost = Rs. 8,180/= + 6 × 100 – 2 × 80

= Rs. 8,620/=

VII. Crash activity 4 – 5

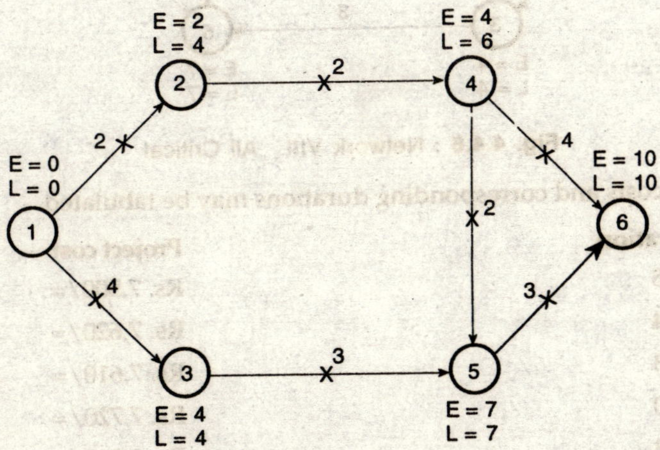

Fig. 4.4.7 : Network VII

Critical path is changed to 1—3—5—6

Duration is 10 days

New project cost = Rs. 8,620/= + 5 × 80 – 4 × 80

= Rs. 8,700/=

Now, if the activity 4 – 6 is decompressed by 2 days (i.e. by the available float), the project completion date of 10 days does not alter, but the cost is reduced 2 × Rs. 60/= = Rs. 120/-

Activity 4 – 5 also can be decompressed by 1 day (1.e. by the available float). By decompressing the activity, the project compeltion date of 10 days does not alter; but the project cost is further reduced by 1 × Rs. 80 /= Rs. 80/=

The entire network is now critical.

The total reduction in cost = Rs. 120/= + Rs. 80/=

= Rs. 200/=

The project cost = Rs. 8,700/= – Rs. 200/=

= Rs. 8,500/=

This is the optimum project cost in minimum time.

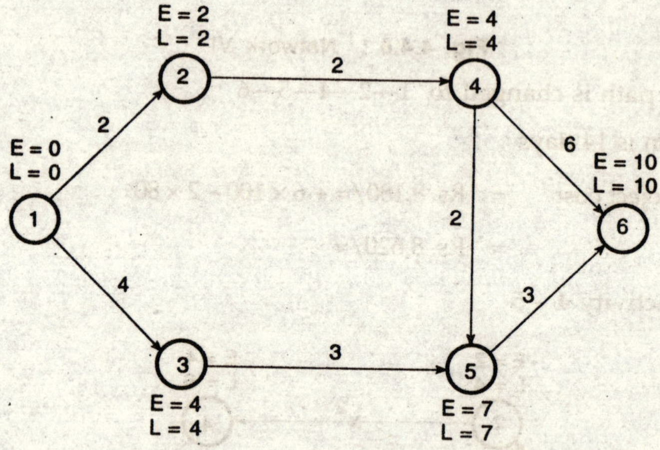

Fig. 4.4.8 : Network VIII. All Critical

The project costs and corresponding durations may be tabulated.

Duration	Project cost
25	Rs. 7,500/=
24	Rs. 7,620/=
23	Rs. 7,610/=
20	Rs. 7,720/=
18	Rs. 8,040/=
16	Rs. 8,180/=
14	Rs. 8,620/=
10	Rs. 8,700/=

The cost of the project may now be plotted against the duration time of the project in figure. It is observed that the overall cost is minimum when the project duration is 10 days. Hence, this is the minimum time for minimum cost.

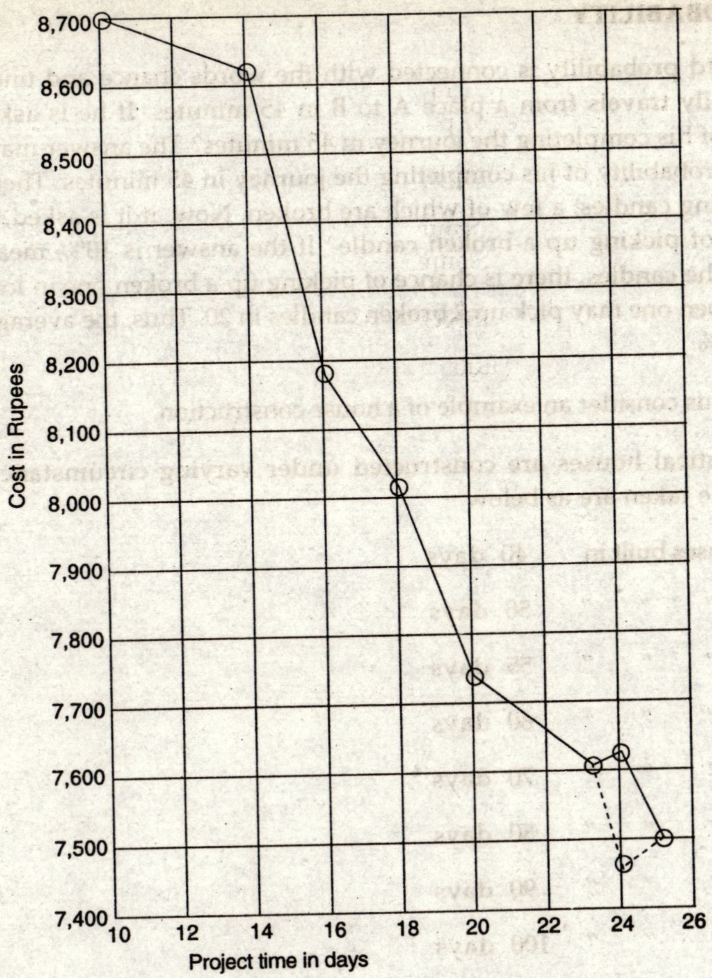

Fig. 4.5 : Graph, Project Duration—Cost

4.5.2 ACHIEVING OPTIMUM PROJECT COST

Referring to Figure 4.4.1 NETWORK I :

Critical path is 1—2 —4 —5—6

Duration is 24 days,

Activity 1 —3, 3 —5, and 4 – 6 are not critical,

Now, if the activity 1 —3 is decompressed by 3 days,

Activities 1—3 —5—6 also becomes critical,

Duration remains 24 days.

Gain in cost is Rs. 50/= × 3 = 150/=

Project cost is Rs. 7,620/= – Rs. 150/=

= Rs. 7,470/=

This is the optimum project cost in optimum time.

4.6.0 PROBABILITY

The word probability is connected with the words chance and uncertainty. A man generally travels from a place A to B in 45 minutes. If he is asked, what is the chance of his completing the journey in 45 minutes? The answer may be 50% or 0.50 is the probability of his completing the journey in 45 minutes. There may be a box containing candles, a few of which are broken. Now, if it is asked, what is the probability of picking up a broken candle? If the answer is 10%, meaning while picking up the candles, there is chance of picking up a broken one in lot of 10, or it may so happen one may pick up 2 broken candles in 20. Thus, the average would be 1 in 10 or 10%.

Now let us consider an example of a house construction.

100 identical houses are constructed under varying circumstances and the duration time taken are as below :

3 houses built in	40 days
10 " " "	50 days
13 " " "	55 days
17 " " "	60 days
20 " " "	70 days
9 " " "	80 days
16 " " "	90 days
12 " " "	100 days

Now, if it is asked what is the probability of building a house in 50 days. The ratio of the number of houses built in 50 days to the total number,

$$\text{Here} = \frac{3+10}{100} = \frac{13}{100} \text{ or } 13\%$$

Similarly, probability of building a house in 70 days will be

$$= \frac{3+10+13+17+20}{100} = 0.63 \text{ or } 63\%$$

and that building a house in 90 days.

$$= \frac{88}{100} = 0.88 \text{ or } 88\%.$$

4.6.1 THE BETA DISTRIBUTION

β—distribution curve fits into the study of probability. Its general characteristics may be used in explaining the time period of the activities.

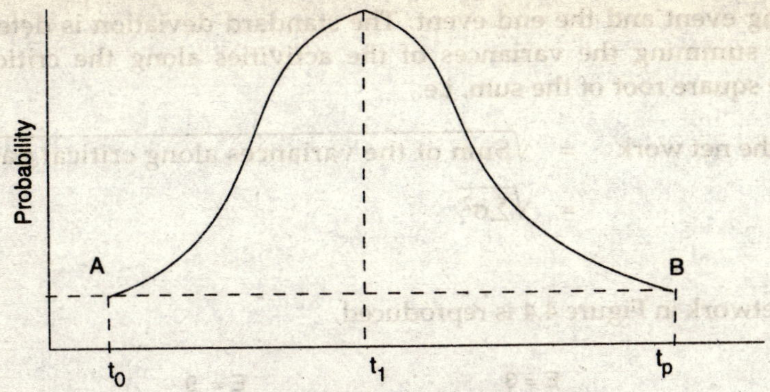

Fig. 4.6 : Time Duration

Point A coincides with optimistic time t_o and point B, with pessimistic time t_p i.e. the range $t_p - t_o$. The node is made to correspond with the most likely time t_1 or t_e.

The standard deviation is approximately, $\sigma = \frac{t_p - t_o}{6}$.

Variance is a measure of dispersion. Since it depends on the range ($t_p - t_o$), larger the variance greater will be the uncertainty.

$$\text{Variance, } \sigma^2 = \left(\frac{t_p - t_o}{6}\right)^2$$

Let us consider the time estimates of activities,

	t_o	t_1	t_p
Activity A	7	9	11
Activity B	4	7	10

Calculation of variance,

$$\text{For A } \sigma^2 = \left(\frac{11-7}{6}\right)^2 = 0.45$$

$$\text{For B } \sigma^2 = \left(\frac{10-4}{6}\right)^2 = 1.00$$

Activity B is more uncertain regarding completion.

4.6.2 NORMAL DISTRIBUTION FUNCTIONS

Normal distribution function gives the values of probability corresponding to the normal deviate, i.e. the distance from the mean expressed in terms of σ.

4.6.3 PROCEDURE OF PROBABILITY CALCULATION

Given a network, the random curve derived from a particular network should be reduced to normal from. T_e values (earliest completin time) of the end event is coincided with the model value. The critical path is the longest path connecting the starting event and the end event. The standard deviation is determined. It is found by summing the variances of the activities along the critical path and taking the square root of the sum, i.e.,

$$\sigma \text{ for the net work} = \sqrt{\text{Sum of the variances along critical path}}$$

$$= \sqrt{\Sigma\sigma^2}$$

Example

The network in Figure 4.4 is reproduced,

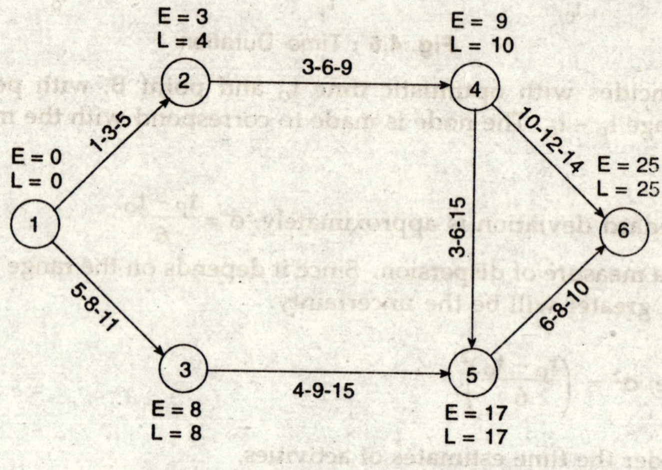

Fig. 4.7 : Network Probability

Critical path is 1—3—5—6

Earliest expected time for completion is 25 days. Let the scheduled completion time be 27 days. Calculation of variances of activities along the critical path

Activity	t_o	t_p	σ^2
1—3	5	11	1.00
3—5	4	15	3.35
5—6	6	10	0.45
			4.80

$$\sigma \text{ for the network} = \sqrt{4.80} = 2.19$$

$$\frac{Ts - Te}{\sigma} = \frac{27 - 25}{2.19} = \frac{+2}{2.19} = +0.91$$

Ts = Scheduled completion time

Te = Earliest completion time

Referring to the Table 4.1, the probability of finishing the project in 27 days is 81.60%.

Table 4.1 Normal distribution function

Normal deviate (–) (1)	Probability (%) (2)	Normal deviate (+) (3)	Probability (%) (4)
0	50.0	0	50.0
– 0.1	46.0	+ 0.1	54.0
– 0.2	42.1	+ 0.2	57.9
– 0.3	38.2	+ 0.3	61.8
– 0.4	34.5	+ 0.4	65.5
– 0.5	30.8	+ 0.5	69.2
– 0.6	27.4	+ 0.6	72.6
– 0.7	24.2	+ 0.7	75.8
– 0.8	21.2	+ 0.8	78.8
– 0.9	18.4	+ 0.9	81.6
– 1.0	15.9	+ 1.0	84.1
– 1.1	13.6	+ 1.1	86.4
– 1.2	11.5	+ 1.2	88.5
– 1.3	9.7	+ 1.3	90.3
– 1.4	8.1	+ 1.4	91.9
– 1.5	6.7	+ 1.5	93.3
– 1.6	5.5	+ 1.6	94.5
– 1.7	4.5	+ 1.7	95.5
– 1.8	3.6	+ 1.8	96.4
– 1.9	2.9	+ 1.9	97.1
– 2.0	2.3	+ 2.0	97.7
– 2.1	1.8	+ 2.1	98.2
– 2.2	1.4	+ 2.2	98.6
– 2.3	1.1	+ 2.3	98.9
– 2.4	0.8	+ 2.4	99.2
– 2.5	0.6	+ 2.5	99.4
– 2.6	0.5	+ 2.6	99.5
– 2.7	0.3	+ 2.7	99.7
– 2.8	0.3	+ 2.8	99.7
– 2.9	0.2	+ 2.9	99.8
– 3.0	0.1	+ 3.0	99.9

4.7.0 CONTROL BY NETWORK/CPM

Network can be used for controlling a project. However, this will not highlight the actual progress of the project. On receiving feed information from the field, the manager shall impose them on the network diagram. The length of line of activity does not indicate graphically the time consumed or the actual progress of the activity. Only the position of the activity in terms of lagging or advancing can be obtained. The manager concentrates on the critical path and notes if the activities on the critical path are progressing as planned. If a critical activity is not complete within the duration period, the critical path may change and some other path may become critical and the total duration period would be changed. In that case, he would have to reschedule the activities and redraw the network. On finding the activity lagging, he may take action to improve its progress and restore the original critical path, if possible. In doing so, he may have to crash the particular activity to shorten the duration period. Obviously, crashing an activity would enhance its completion but would involve additional cost which the client may not like to incur. Hence, the manager shall have to decide whether to allow the overrun of time causing consequent delay in completion of the project or to crash the activity or the activities to restore the original work programme. Financial implication of both the alternatives shall have to be evaluated and decision taken in consultation with the client. Thus network with CPM can be effectively used for controlling a project.

As mentioned earlier, a master network may be divided into sub-networks of important and complete items. Considering these as activities, master network can be drawn. In such case, monitoring shall have to be exercised on each sub-network and then the master network. Financial implication in both the cases shall have to be evaluated.

4.7.1 DIFFERENCE BETWEEN PERT AND CPM

	PERT	CPM
i	It is basically analytic in concept and designed as probabilistic network.	It is basically synthesising in concept and is termed as deterministic network.
ii.	It is formed of event oriented diagrams.	It is formed of activity oriented diagrams.
iii.	The time estimation of duration is based on three probable times.	Time estimation of duration is based on experience of similar activity without having any allowance for uncertainty.
iv.	Duration times of activities are not related to cost.	Duration times are related to cost.
v.	The method is generally used for Research and Development projects, Production development, and similar projects involving uncertain factors.	The method can gainfully be used for repetitive work – construction activities, maintenance and repairs and production control, planning and scheduling.

4.7.2. DIFFERENCES BETWEEN NETWORK AND BAR-CHART

	Network	Bar Chart
i.	Logical planning of activities can be separated from time scheduling;	Planning cannot be separated from scheduling ;
ii.	Interrelationship of activities is established through out the project;	Very little such relationship is available;
iii.	Effect of any delay or change is readily available;	Generally a few immediate effects can be identified;
iv.	Time period may have to be reallocated at any point, but redrawing of the network is rarely required;	Frequently becomes completely out of schedule necessitating redrawal of chart;
v.	Resource requirements with time are always known in advance;	Very limited indication is available in this respect;
vi.	Spare times are available e.g. slack, floats etc. systematically calculated and allocated;	There is no spare time in the planning ;
vii.	Critical activities and bottlenecks are highlighted;	No such identification of critical activity; consequently crash resources are often wasted on non-critical activities;
viii.	It helps in maintaining the time schedule as problems can be foreseen and decisions can be taken adequatèly in advance;	Last minute or late knowledge of problem often upsets target;
ix.	Overall costs can be minimised by constant review and curtailment of overhead cost by decompressing the non-critical activities;	No such possibility exists;

4.7.3 ADVANTAGES OF NETWORK ANALYSIS

Various advantages can be derived by application of network analysis in capital projects in their execution starting from *ab initio* to completion stage.

i. It limits possibility of overlooking any activity in a project;

ii. It shows interrelationship of all the activities in a project;

iii. It shows clearly the relationship controlling the order of performance of various activities than any other management tool;

iv. It helps quick and easy communication of ideas. The pictorial approach helps much than any verbal instruction;

v. It contributes disciplined thinking. It trends to prevent error, commission caused due to careless planning and generates confidence in the integrity of the schedule;

vi. It provides time schedule containing much more information than any other technique;

vii. It identifies those jobs which are critical in a project. Attention can be put on them.

viii. It predicts trouble areas, showing the timing and warns against any future problem.

ix. It gives an accurate forecast of resources requirements – materials/labour, quantity and variety.

x. It provides method of resource allocation suiting the condition and to maintain or to minimise overall cost;

xi. It integrates all elements of a programme in detail as dsired by the management;

xii. It relates time to cost, which allows regulating any proposed change.

xiii. It facilitates reporting and project documentation;

xiv. It simplifies communication and improves co-ordination through the use of detailed common language i.e. action plan.

4.7.4 LIMITATIONS OF NETWORK

Following are the main limitations of Network.

i. Network having large number of activities, if analysed properly and in details, require large number of mandays for calculations, which is a very monotonous job.

ii. Repetitive manual calculations are seldom error-free and an incorrected early error can produce an useless programme.

iii. It is very difficult to identify the error very precisely enabling correction without undue delay, resulting wastage of analysing efforts.

iv. It is often necessary and advantageous to analyse different sequences present in large networks for easy understanding. But this is not possible in a limited time in this method.

v. It is often difficult to carry out extensive checks on the data supplied and the validity of the network due to its complexity.

4.7.5 APPLICATION OF NETWORK ANALYSIS

Construction

One of the most widely accepted application of network analysis is in the area of construction activities in all the disciplines e.g. construction of buildings of any magnitude, cinema houses, stadium, cultural halls, dams, roads, bridges etc.

Repairs and maintenance

In the field of Rapairs and Rehabilitation of structures, the technique can be effectively applied. In planned maintenance and preventive maintenance, the conventional approach is mostly found to be quite inadequate in consideration of the narrow time span and limited resource often available. The need for skilled maintenance and manpower necessary can be effectively met by the network analysis for economy in cost and time.

Chapter 5

Controls and Progress Report

5.1.0 CONTROL

Bar-Chart, CPM, PERT etc. provide for ample control of a project during its execution. Apart from control of execution of work, other controls are necessary for efficient management of the project.

5.2.0 QUALITY CONTROL

Quality control is one of the most important activities of management. Progress of a work without quality control cannot be considered as a work at all. It is the foremost duty of the manager to maintain the quality throughout the duration of the work. This he would do with the help of efficient, experienced and well conversant group of field staff and also with various testing machines and implements. Control of quality shall be exercised on the daily execution of different items of work, specially those which would be covered up.

5.2.1 QUALITY OF MATERIALS

Before using any material in work, the same should be inspected thoroughly to ensure that the material is of quality as specified. If necessary, it may be tested at site or in laboratory.

Cement to be used shall be of fresh stock. This may be tested at site. Cement shall be bluish grey cement colour and on pushing hand inside a bag full of cement, one shall have cool feeling. It may be tested by making briquettes and breaking those after 7 days. These tests do not give very correct results, but are indicative. In case of any doubt, cement sample shall be sent to laboratory for testing its physical and chemical properties.

During concreting, concrete cubes shall be taken random from mixed concrete for testing compressive strength after 7 days and 28 days. Concrete is generally poured for long durations. Throughout the operation, its consistency shall be maintained ensuring the water–cement ratio. This can be done easily by slump test with the help of a slump cone.

The materials like aggregates shall be clean, sharp, hard, free from any dirt or dust. If necessary, those shall be washed and cleaned.

Bricks used shall be of first class quality, shall emit ringing sound when struck together or by observing its breaking pattern from a fall which would indicate the quality. Before use the bricks shall be soaked in water at least for 12 hours.

Timber used shall be checked thoroughly to ensure that it is free from knots, damaged fibre, diseases like rot, fungi attack etc. Timber shall be well seasoned and treated.

Paint or varnish to be used shall be from sealed containers, shall be used properly with brush or spray guns without adding any thinner.

Materials required and used in Civil Engineering works are of different varieties and in large quantities. These are generally handled by human being. Hence, it is difficult to control the quality unless watchful eyes are kept on them.

5.2.2 QUALITY OF WORKMANSHIP

Workmanship is again another field of numerous problems and loopholes. Workmanship shall be maintained for quality control. The finishing of the work shall be of A grade. The slope, line, level shall be maintained as specified in the drawing. Possibility of human error and mistakes leading to defects in workmanship are there as human factor exists almost throughout the entire operation. However, careful and watchful manager can still deliver good end production by exerting his control.

Progress and quality control shall proceed side by side. One should not pose problem to other.

5.3.0 FINANCIAL CONTROL

This is another field where control of the management is called upon. Responsibility of the management does not end with planning and execution of the work. Management includes financial planning and control during the entire process of execution, so that the operation ends successfully without any fund problem which would otherwise hamper the physical progress. Physical progress of a project cannot be ensured without proper financial inflow of fund required at different stages as planned. This part of the management is better controlled by a Cash-Flow Chart. The cash-flow chart will indicate the flow of fund required, expenditure, and return, if any expected from the project at different stages. It shows the expected financial benefit that may accrue at the completion of the project.

EXAMPLE

A project costing Rs. 500 lacs is proposed to be financed by borrowing from a financial institution. Loan will be drawn as required. The borrowed amount will bear an interest at the rate of 10% per annum. The project is expected to yield Rs. 700 lacs in a span of 8 years starting from the third year.

Draw a Cash-Flow Chart.

CASH-FLOW CHART All figures are in Rs. lac

Year	Expenditure during the year	Loan during the year	Return from the project	Return of loan with interest	Interest on loan during the year 10%	Total loan & interest at the end of the year cumulative	Remarks
1st	50	50	—	—	2.50	52.50	
2nd	100	100	—	—	10.25	162.75	
3rd	100	50	50	—	18.78	231.53	
4th	100	—	100	—	23.15	254.68	
5th	100	—	100	—	25.47	280.15	
6th	50	—	100	50	28.02	258.17	
7th	—	—	100	100	25.82	183.99	
8th	—	—	100	100	18.40	102.39	
9th	—	—	100	100	10.24	12.63	
10th	—	—	50	13.89	1.26	36.11	Surplus
	500	200	700	363.89			

i. It would be seen from the Cash-Flow that the project was self-financing.

ii. Borrowed amount would be returned fully with interest in 10 years.

iii. The project would yield Rs. 36.11 Lac as surplus after meeting all the liabilities.

If the actual financial inflow from the project i.e. return from the project goes wrong, the net surplus at the end of 10th year Rs. 36.11 lac would be eroded partly or fully. In the event of such occurrence, the manager foreseeing that the final yield might be eroded and/or there is possibility of loss, he might suggest ways and means for enhancing the return from the project much ahead to counter the possibility of any loss in the future and keep the project financially viable.

5.4.0 PROGRESS REPORT

For easy and quick understanding of the status of a project, a periodic progress report need be prepared and submitted to the Client or the Authority for his appraisal. The report should be simple and easily understandable without any technicality.

5.4.1 PERCENTAGE PROGRESS

Percentage progress report system has been evolved to satisfy the requirement. The Bar-Chart, Network Analysis, CPM are the tools available for the management for better management of the projects and exerting control at various level of the project. But none of the methods can indicate the overall progress of the project which is often necessary for monitoring both financial and physical progress of the project. Client or the authority generally like to know the status of the project at any particular point of time or at regular intervals and that too in a simple way. A percentage progress report would provide a satisfactory communication from the management of the project.

The percentage progress is a method of exercising overall control of a project as it indicates the percentage of work already done and the percentage of work remaining. It is a simple method and can be understood easily. But it does not help much in indicating the particular activity causing any delay in the process, which is better indicated in network. Process depends upon assessment of the physical progress of the elements of the work at site, which is collected on approximate estimation basis by the field staff and is subjected to error. However, the method extends help in ascertaining the overall progress of a project.

In the method, the project is divided into a possible number of sub-heads which can be identified individually. The financial weightage of the sub-heads in relation to the whole project is noted on each sub-head in separate proforma. A project may comprise of many sub-heads.

EXAMPLE

Construction of Health Centre Rs. 201 lac

	Sub-heads	**Cost**	**Weightage**
i.	Construction of hospital building	Rs. 106 lac	0.53
ii.	Construction of staff qrs.	Rs. 50 lac	0.25
iii.	Construction of compound wall	Rs. 10 lac	0.05
iv.	Construction of outhouse	Rs. 5 lac	0.02
v.	Sinking or deep tubewell and providing water supply	Rs. 15 lac	0.075
vi.	Sewerage and drainage	Rs. 15 lac	0.075
			1.00

Separate sheets are now developed for each sub-head showing all detailed items of work and percentage of individual item is calculated on the basis of cost of the sub-heads. On the basis of the percentage of the items, the weightage of the items in relation to the cost of the sub-head is calculated. The weightages are added and then multiplied by the weightage of the sub-head to give the percentage of the work done of the sub-head in relation to the total project.

The percentage thus obtained are added to find out the percentage progress of the whole project.

In this method, one can find the following :

i. Percentage progress of the total project,

ii. Percentage progress of each sub-head, and

iii. Percentage progress of each item belonging to a sub-head.

Thus the method reveals more clearly the progress of the work in terms of financial percentage progress achieved.

The method indicates a realistic index of progress of a project; but it involves a lot of field work for collecting the information of each item of work scattered all over the project site and may be at different places away from each other. Again, the method do not indicate the situation causing any problem or the activity responsible for any delay. It does not indicate any cause for delay which need correction. However, the client likes to know what percentage of the work has been completed and he may like to calculate himself the time that would be required for completion of the project. He also can have an idea of the fund position.

The method can be utilised by considering physical progress instead of financial progress by considering physical weightage in place of financial weightage. But this would require very careful judgement. For example, earthwork in excavation in foundation trenches may have financial weightage of 0.1% but its physical weightage may be 5%. Again this may vary according to project or its site of work.

Specimen of proforma for reporting percentage progress is given.

PROGRESS REPORT

Proforma I

Name of the project — Construction of Health Center at

....

Progress up to — ....

Activities	Weightage	Percentage progress	Percentage progress with weightage
I. Construction of hospital building.	0.53		
II. Construction of Staff Qrs.	0.25		
III. Construction of compound wall and internal roads and pathways	0.05		
IV. Construction of out house	0.02		
V. Construction of deep tubewell and external plumbing works	0.075		
VI. Sewerage and drainage	0.075		
	1.00		

Component I

Construction of Hospital Building Weightage 0.53

Activities	Weightage	Percentage progress	Percentage progress with weightage
1. Earthwork in excavation	0.01		
2. Foundation concrete	0.04		
3. Brickwork upto plinth	0.04		
4. Brickwork ground floor	0.12		
5. Gr. Floor roof slab	0.14		
6. Brickwork first floor	0.12		
7. First floor slab	0.14		
8. Rendering internal	0.06		
9. Rendering outside	0.06		
10. Fixing door and window frame	0.05		
11. Fixing door & window shutters	0.05		
12. Electrical installations	0.05		
13. Sanitary and plumbing installations	0.05		
14. Finishing works	0.05		
15. Site clearance	0.02		
	1.00		

Component II

Construction of Staff Quarters Weightage 0.25

Activities	Weightage	Percentage progress	Percentage progress with weightage
1. Construction of foundation and plinth.	0.20		
2. Ground Floor brickwork	0.12		
3. Ground floor roof slab	0.12		
4. Brickwork first floor	0.12		
5. First floor slab	0.12		
6. Rendering internal	0.05		
7. Rendering outside	0.05		
8. Fixing door and window frames and shutters	0.10		

Contd.

Activities	Weightage	Percentage progress	Percentage progress with weightage
9. Electrical installations	0.05		
10. Sanitary and plumbing installations	0.05		
11. Finishing works	0.02		
	1.00		

Component III

Construction of Compound Wall and Internal roads and pathways — Weightage 0.05

Activities	Weightage	Percentage progress	Percentage progress with weightage
1. Construction of foundation	0.15		
2. Brickwork above ground level	0.40		
3. Rendering and finishing works of wall	0.10		
4. Laying of pathways	0.15		
5. Laying of internal roads	0.15		
6. Finishing works including site clearance	0.05		
	1.00		

Component IV

Construction of Outhouse — Weightage 0.02

Activities	Weightage	Percentage progress	Percentage progress with weightage
1. Construction of foundation and plinth	0.25		
2. Brickwork in superstructure	0.30		
3. Roof slab	0.25		
4. Fixing of door and window shutters	0.10		
5. Finishing works including site clearance	0.10		
	1.00		

Component V

Construction of deep tubewell and external plumbing works

Weightage 0.075

Activities	Weightage	Percentage progress	Percentage progress with weightage
1. Moblization at site	0.05		
2. Drilling of tubewell	0.20		
3. Lowering of pipes	0.40		
4. Testing of tubewell	0.05		
5. Laying of external water supply distribution network including construction of pump house and pump	0.20		
6. Construction of water supply lines and testing	0.10		
	1.00		

Component VI

Sewerage and drainage

Weightage 0.075

Activities	Weightage	Percentage progress	Percentage progress with weightage
1. Excavation of trenches	0.25		
2. Laying of sewerage pipes	0.50		
3. Testing	0.05		
4. Drainage works	0.20		
	1.00		

Chapter 6

Safety and Precautions

6.1.0 SAFETY AND PRECAUTIONS

In management of any construction work, safety and precautionary measures need be given priority to avoid future bottleneck of the progress due to accident causing damage to human lives and/or property.

Safety measures would be preventive as well as follow up actions as may be required after any mishap.

Preventive Measures

i. **Plan showing underground services**

Plan showing the existing underground services like electric cables water supply and sewerage lines, telephone lines, etc. should be prepared much ahead of commencing the actual work at site for guidance during execution.

ii. **Barricade**

The site should be protected by erecting barricade.

iii. **Survey of the neighbourhood**

The site and the neighbourhood shall be properly surveyed before commencement of the work, specially before starting any digging work for detecting if there be any existing foundation or any other structure.

iv. **Temporary supporting structures**

Temporary supporting structures, e.g. shoring, timbering of trenches, staging, shuttering, propping, etc. shall be strong enough, well designed and erected properly.

v. **Ladder**

Ladder used for climbing to heights shall be checked prior to putting into service. It shall be kept at proper angle and held by a man at the ground.

vi. **Working platform**

The working platform on the staging should be strong and provided with guard board.

vii. **Safety net**

When the work is at a higher level, proper safety nets should be provided around the working place to protect the labour who might have slipped.

viii. **Plants and machineries**

Thorough checking of plants and machineries shall be undertaken before putting them into service.

ix. **Trained labour for specialised work**

Only trained labours shall be engaged for specialised works.

x. **Training of labours**

In general, labours engaged shall be trained for the job.

xi. **Alcoholic and drug addict labour**

Alcoholic and drug addict labours or staff should not be allowed to enter the site.

xii. **Handicapped labour**

Handicapped and child labour should be engaged in specific jobs suitable for them.

xiii. **Head gear**

Wearing of head gear or helmet should be made compulsory for all labour, staff, supervisors and visitors. Persons engaged in painting work shall be provided with mask and goggles, and if necessary with respirators.

xiv. **Underground services**

No digging work or piling work shall be allowed without checking properly and ensuring that there is no underground services below.

xv. **Illumination**

The project site shall be properly illuminated, specially if any night work is envisaged.

xvi. **Caution board**

Caution and warning signs shall be displayed at prominent places and shall be made to understand by all the staff and labours.

xvii. **First aid and medical help**

A medical and first aid box must be kept at site in any easily accessible place and the same must be known to the staff engaged in the work.

The name and address, location and telephone number of the nearest medical center or hospital should be known and the same should be displayed in a board at a prominent place.

The telephone numbers of the following should be kept in display at a prominent place.

— Police Station

— Fire Station and

— Ambulance.

xviii. **Negligence and complacency**

This is the most damaging cause and is the root of most of the accidents at site. The manager should keep everybody alert and shall not encourage any negligence or complacency. He himself should not indulge in uttering any loose word against taking any precautionary measure in presence of subordinate staff or labour.

6.1.1 FOLLOW–UP MEASURES

In spite of all the precautionary measures, if there by any mishap at the site causing loss or damage to any person or property all sorts of help should be extended for providing immediate relief and then for proper treatment or restoration of the damage.

6.2.0 LITIGATION

In most of the projects, the management faces litigation. With the increase in knowledge of democratic rights and legal protection this would increase. Litigation may be started by public, employees or labour. Management should keep watchful eye and try to avoid litigation. Litigation means hindrance to progress of work. Litigation may cause suspension of the work for days, months and even for years.

It would cause financial loss and might damage the project permanently.

6.3.0 CHECK/SNAG LIST

When the contractor reports that the project is complete, the manager shall ask his people to go round the project and inspect in details for checking if the works are really complete and simultaneously record if there is defect and prepare a check/snag list. The inspection shall be made quickly and the complete snag list shall be prepared at different stages of the work. The manager after going through the list shall pass on the same to the contractor with his remarks for mending and making good the snags. When the snags are removed, the work shall be inspected again. A completion certificate shall be issued by the Engineer-in-Charge on completion of the project including cleaning of the site and ensuring that there is no liability due from the contractor. Sample of Check/Snag List is enclosed in Annexure I.

6.4.0 PROBLEMS RELATED TO EXECUTION OF PROJECT

During execution of a project problem may arise which the management might not have foreseen prior to commencement of the work. Management shall take a note of the problem and solve it for maintaining the progress and achieving the goal of completing the project in schedule time at optimum cost.

Problem arising during execution of a project may be various and of different kinds. It is not possible to make out a list of probable problems. A problem has to be studied and then dealt with separately according to its nature and magnitude. Problems may be mainly of two kinds.

i. **Problem due to defective, inadequate or vague provision in the tender clauses.**

Some of the probable problems may be as below :

a. Inadequacy in provision of tender clauses may encourage the contractor to raise claim for extra payment. Proper care shall be taken while drafting the tender document so that no such problem can arise during execution. However, when such case has arisen, the claim shall be examined carefully and if it is found genuine, should be admitted for early solution of the problem.

b. Problem may arise due to provision of insufficient duration time for the work. Time provided should be carefully calculated taking into account of all the activities correctly.

Contractor's claim for extra time should be carefully examined and if found proper, the problem should be solved by extending the time period.

c. Non-provision of clause for escalation of price of materials and labour may lead to serious claim and create problem subsequently. Clause for escalation of prices should be included if the project is likely to continue for more than six months.

When such problem arises, it should be solved by mutually agreeing to the formula based on Price Index of the Reserve Bank of India.

ii. **Construction problem arising during execution of the project**

In this category many unforeseen problems may arise. Some of the problems with their solutions are given below:

a. **Delay in handing over the site**

Due to delay in handing over of the site of the project to the contractor, there may be problem as starting of the project would be delayed. Delay may also occur due to non-removal of encumbrances in the land.

These problems shall be solved by extending the date of completion if the contractor agrees, and due to extension of the period price escalation does not create further problem. If the extension invokes escalation of prices, price escalation clause shall have to be taken into consideration.

b. **Soil investigation not done**

The land for execution of the project might appear to be good and level, suitable for construction; but after digging the foundation it may be seen that the soil below the proposed foundation bed was decomposed and compressible. Foundation could not be laid on such soil. Such situation would not have arisen if proper soil investigation was done during planning to ensure the soil condition. Such case would be embarrassing and give rise to problem. Decision may have to be taken for removing the compressible soil and take the foundation deeper and lay it on virgin soil. Alternatively, after excavating and removing the soil, the trench may have to be filled with sand and lay the foundation at the desired level. All these would cause problem and the management would be required to solve the problem quickly and efficiently.

c. **Soil below the foundation was not consolidated**

The top 1.5 m (approximately) soil might be good, sandy silt having satisfactory bearing capacity. But the soil below was almost in semi-liquid condition having very low density and poor bearing capacity, 2.5 tonnes per sq.m.

If the foundation was laid at a level 1.2 m below ground level, the pressure bulb would be in the zone of soil having very poor bearing capacity. The foundation would be susceptible to settlement. The building proposed to be constructed was of brick masonry with ordinary brick foundation on cement concrete (1 : 3 : 6) bed. This was a problem requiring redesigning of the foundation.

The foundation was raised at higher level, 0.80 m below the ground level. Masonry foundation was replaced by R.C.C. flexible strip raft having a beam on the raft slab on which the masonry wall was built. The raising of the foundation would raise the pressure bulb and the intensity of pressure would be much less in the zone of the weaker soil. Thus the foundation would be less susceptible to settlement.

d. **Foundation on expansive soil**

After excavation it was observed that the soil on which the foundation was to be laid was expansive soil or black cotton soil. This type of soil when saturated with moisture becomes very soft, have very poor bearing capacity and expands in volume. Due to increase in volume, it exerts upward thrust on the foundation. Cracks appear in the wall. The nature of the cracks would be from the foundation upwards, very fine at the lower level and wider in the upper region.

The effective solution of the problem would be to construct the foundation on under-reamed piles. These piles would provide increased bearing area for distribution of pressure on the soil. In moist condition, due to expansive character, the soil would exert upward pressure and try to push the foundation upwards. The reams would hold the foundation in position.

e. **Unforeseen problem**

A deep foundation having basement was under construction very close to a single-storey building with isolated footing and columns. The excavation was touching the edge of the isolated footing of the adjoining building.

The excavation was 5.3 m deep while the isolated footing was at a depth of 1.8 m from the ground level.

The soil was not black cotton, but was blackish clay. It was hard like stone when dry, but would be very soft when moist. For protection of the vertical walls of the excavation, close sal bullah piling was done all around. Excavation was done like a pond at the central section and R.C.C. raft with columns in location was laid. The vertical R.C.C. walls at the edges were taken up in stretches of 6.0 m in length at a time. While excavating this portion, the vertical earth was protected by close sal bullah pile walling. The walling was supported by three rows of inclined supports—one at the top, one at two-third height and the third one at one-third height. The tops of the inclined supports were fixed on the horizontal runners and the toes were held by longitudinally laid sal bullah fixed on the R.C.C. raft by steel pegs.

The system was working well and the work was progressing as per schedule. The whole operation was scheduled to be complete before the monsoon expected by 8th of June. Just before the last two stretches, an accident occurred at the site. A boy was killed and a man was seriously injured. Both were outsiders. This upset the whole programme. The work was suspended for about a month.

When the work could be resumed, monsoon has already set in. But the incomplete work could not be left in that condition which was already threatening the adjoining building.

The entire working area was covered by erecting tarpaulin shed and the work continued in spite of the monsoon. Quick-setting compound was used for expediting the concrete works.

Though the problem was mainly due to working under monsoon condition with restricted working place having a busy and occupied building adjoining, the real problem was seepage of water through the close sal bullah walling. Seepage of water only was not a problem, rather would have been helpful as lateral water pressure would have been reduced. But, soil was flowing with water which created real problem. This flow of soil from below the footings of the adjoining structure could endanger the building. However, with careful and constant watch and taking precautionary measures, the seepage was controlled and the work was completed. Seepage was controlled mainly by inserting jute packing in between the sal bullahs, where found necessary.

In spite of all expediting methods adopted, 30 days normal time work could be completed in 20 days, working 12 hours per day. Due to crashing of the activities, the cost of 12.0 m length of wall was increased by about 50% than the normal cost.

f. **Flowing of soil**

During execution, in places where ground water level is high or in monsoon condition, dewatering the foundation by pumping may be required for excavation of the foundation trench or concreting in foundation. During dewatering, so long only water is pumped, there is no problem, or danger; but as soon as soil starts flowing with water as mud or slash, there is possiblity of danger of the adjoining structure, if there be any.

A case was observed when foundation trench of a building was being dewatered and soil was flowing from the adjoining area. A partly constructed building was close to it. Soil from below the partly constructed building flowed and partial void was created in the region below the foundation.

The phenomenon was not noticed during the construction. Both the buildings were complete and finished. While clearing the site and levelling the ground, it was observed that the first building had tilted and there was differential settlement, but it had not suffered any distress. The second building was unaffected.

Investigation was undertaken. It was found that soil below the titled building had stabilised and there was no further danger of the building. Fortunately, the building could be declared safe.

Carelessness and lack of observation during execution was responsible for such occurrence which was not desired.

There may be many such cases with problem. The management should keep vigil on the proceedings at site during construction with the ultimate aim of completing the project in minimum time and optimum cost.

Chapter 7

Economics of Buildings

7.1.0 STUDY OF ECONOMICS OF A BUILDING

Before launching a Building Project, the manager of the project should examine the project in all aspects, e.g. planning, design, construction technology, materials to be used and economics of the building prior to deciding upon the management technique to be adopted. This detail study of the project is necessary to ensure completion of the project without wastage of time, materials, manpower and finance. For optimising economy, attention must be given from the planning and design stage. Deciding upon the shape and size of the building is important as these influence the cost of the structure.

7.1.1 CONSTRUCTION TECHNOLOGY

Construction technology adopted is still the old traditional one. The technology is selected in consideration of suiting the masons, carpenters and other tradesmen. The trades have passed from generation to generation without adoption of any innovation and idea. The tradesmen are averse to adoption of any new technology.

The manager should give serious thought over all these and should not hesitate to adopt new technology, new substitute materials to achieve his goal. In doing so, he will have to examine the scheme and shall have to evaluate the cost of the building as accurately as possible.

7.2.0 ASSESSMENT OF QUANTITY OF MATERIALS AND LABOURS

Assessment of quantity of different materials and labours of different categories is generally made on percentage basis on the total cost of construction. But, it is a very approximate method and differs widely as cost of construction differs widely from place to place and the price of materials and labour do not follow any rationality and do not bear any proportion.

To obviate the difficulty and to arrive at reasonably correct quantities of different materials and number of different categories of labour required, Central Building Research Institute has worked out certain statistical equations for building portion, which relate to consumption of each material/labour to plinth area of the building. For services portion, the norms/requirements have been given in tabular form. Statistical relations have been worked out for residential and office buildings – both load bearing and framed structure.

Statistical relations are furnished in tabular form. These are based on specifications of the structures.

TABLE 7.1

Specifications adopted in residential building

Element	Single and double storey construction	Four storey (RCC Framed Building)
1	2	3
Foundation	Plain cement concrete foundation 1 : 5 : 10 with stone ballast, brickwork in cement mortar 1 : 6, 38 mm thick DPC with cement concrete 1 : 2 : 4 and a bitumen coating on top @ 1.7 kg/sqm, sand filling 100 mm in plinth.	RCC 1 : 2 : 4 column footing on lean concrete (1 : 4 : 8) mat with stone ballast, RCC 1 : 2 : 4 plinth beams finished smooth with 6 mm cement mortar 1 : 3, brickwork in cement mortar 1 : 6, sand filling 100 mm in plinth.
Frame		RCC 1 : 2 : 4 columns and beams finished smooth with 6 mm mortar 1 : 3.
Walling	Brick work in cement mortar 1 : 6, load bearing walls one brick wide and partition walls half brick thick, RCC work 1 : 2 : 4 in lintels, beams and chajjas finished smooth with 6 mm thick cement mortar 1 : 3.	Brick work in cement mortar 1 : 6 RCC work 1 : 2 : 4 in lintels, becams and chajjas finished smooth with 6 mm thick cement mortar 1 : 3.
Floor	38 mm thick cement concrete 1 : 2 : 4 laid over cement concrete 115 mm thick 1 : 5 : 10 with stone ballast.	38 mm thick cement concrete 1 : 2 : 4 laid over 115 mm thick cement concrete 1 : 5 : 10 with stone ballast in ground floor and over RCC slab in upper floors.
Roof	RCC slab with lime terracing avg. 110 mm thick with bitumen coating underneath @ 1.7 kg/Sqm.	RCC slab with lime terracing av. 110 mm thick with bitumen coating underneath @ 1.7 kg/sqm. and top finished with 40 mm cement concrete. Brick work in cement mortar 1 : 6 in parapet finished with cement concrete 1 : 2 : 4 coping.
Joinery	Timber frames 100 × 75 mm, 38 mm thick panelled shutters for doors and fully glazed shutters for windows.	Timber frames 100 × 75 mm, 38 mm thick timber panelled shutters for doors and fully glazed shutters for windows.

Element	Single and double storey construction	Four storey (RCC Framed Building)
1	2	3
Finishes	Sand and cement plaster on the wall 1 : 6, 12 mm thick on smooth sides and 20mm thick on rough sides, cement plaster 1 : 3 in skirting and dado, internally white washed, externally colour washed, painting wood-work with oil paint.	12 mm thick cement plaster 1 : 4 on internal faces of the walls, and 20mm thick cement plaster 1 : 3 on external faces of walls including neeru finish, internally white washed and externally colour washed, painting wood-work with oil paint.
Stair	RCC stairs in double storey.	RCC stair.
Fittings & Fixtures	Precast RCC shelves, M.S. round bar in window grating, raised cooking platform.	Precast RCC 1 : 2 : 4 shelves, M. S. round bars in window gratings, timber cupboards, raised cooking platform.

TABLE 7.2

Equations for materials requirement (A is the plinth area in m^2) in residential buildings

	Equations		
	Single Storey	Double Storey	Four Storey * (RCC Framed)
1	2	3	4
Bricks (% nos.)	$2.26A + 66.8$	$2.15A + 63$	$-26.2 + 2.56A - 0.0096A^2$
Cement (tonne)	$0.153A + 0.57$	$0.145A + 0.54$	$0.182A - 0.35$
Steel (kg)	$21.3A - 314$	$21.97A - 305$	$-1491 + 92.0A - 0.36A^2$
Sand (m^3)	$0.47A - 7$	$0.43A - 5.6$	$0.361A - 0.38$
Coarse aggregate :			
(i) 20 mm down (m^3)	$0.176A - 0.21$	$0.178A - 0.21$	$0.295A - 0.75$
(ii) 40 mm down (m^3)	$0.145A + 1.5$	$0.075A + 0.78$	$0.45 + 0.0027A + 0.0001A^2$
Brick aggregate (m^3)	$0.113A - 0.83$	$0.056A - 0.42$	$0.021A + 0.01$
Timber frames and shutters (m^3)	$0.019A + 0.23$	$0.019A + 0.23$	$0.02A + 0.11$
Timber shuttering (m^3)	$0.0042A$	$0.0042A$	$0.009A - 0.03$
Ballies (m)	$0.504A$	$0.504A$	$0.90A + 2.41$
Lime (quintal)	$0.145A - 0.35$	$0.083A - 0.17$	$0.063A - 0.08$

	Equations		
	Single Storey	Double Storey	Four Storey * (RCC Framed)
1	2	3	4
Surkhi (m^3)	0.052A – 0.37	0.026A – 0.18	0.01A
Glass (m^2)	0.047A	0.047A	0.047A
Bitumen (kg.)	1.836A – 9	0.918A – 4	0.357A + 0.14
Primer (lt)	0.048A	0.048A	0.045A + 0.56
Paint (lt)	0.08A + 0.27	0.08A + 0.27	0.075A + 0.93

* Equations for superstructure only. For sub-structure, refer Table – 7.4.

TABLE 7.3

Equations for labour requirement (A is the plinth area in m^2) in residential buildings

Labour	Equations		
	Single Storey	Double Storey	Four Storey * (RCC Framed)
1	2	3	4
Mason (days)	1.335A + 28	1.355A + 6	1.57A – 2
Carpenter (days)	1.184A – 9	1.194A – 9	1.61A
Painter (days)	0.19A	0.19A	0.19A
Blacksmith (days)	0.269A – 4	0.274A – 1.4	–16 + 1.01A – 0.004A^2
Mazdoor (days)	4.769A + 32	4.91A + 13	5.49A – 9.2

* Equations for super-structure only. For sub-structure, refer Table 7.4

TABLE 7.4

Equations for Materials and labour requirement of sub-structure in 4–storied (RCC Framed) Residential Building

Materials/Labour	Equations
Materials	
Cement (tonne)	0.0204 A – 0.014
Sand (m^3)	0.036A
Coarse Aggregate 20 mm and down (m^3)	0.071 A – 0.01
Steel (kg)	–171 + 10.46A – 0.04 1A^2
Timber shuttering (m^2)	0.0007 A
Ballies (m)	0.036 A + 0.06
Labour	
Mason (days)	0.023 A
Carpenter (days)	0.05A
Blacksmith (days)	–1.6 + 0.1A – 0.0003 A^2
Mazdoor (days)	0.343 A

TABLE 7.5

Equations for Materials/Labour requirements for Office Buildings

Materials/Labour	Equations
Cement (tonne)	0.1925 A + 18.52
Fine sand (m^3)	0.08 A + 105.5
Coarse sand (m^3)	0.2592 A – 80.94
Coarse aggregate (m^3) 20 mm nominal size	0.2728 A – 48.5
Coarse aggregate (m^3) 10 mm nominal size	0.1164 A – 20.74
Coarse aggregate (m^3) 40 mm nominal size	0.0151 A + 73.91
Brick ballast 25 mm nominal size (m^3)	0.0426 A – 38.37
Timber for form work (m^3)	0.005 A + 11.19
Timber for joinery (m^3)	0.0024 A – 0.53
Ballies (m)	0.5507 A + 797.5
Bricks % nos.	1.1829 A – 524.23
Deformed bars (tonne)	0.0479 A
Door shutters (m^2)	0.0636 A – 17.07
Window shutters (m^2)	0.1117 A + 93.26
Glass (m^2)	0.1407 A + 55.99
Primer (lt)	0.0256 A + 9.7
Paint (lt)	0.0322 A + 7.24
Lime qtl	0.0754A – 51.21
Surkhi m^3	0.0204A – 18.39
Marble Chips qtl	0.1338A – 48.52
Marble powder m^3	0.0012A – 0.36
Mason (day)	1.1314 A – 407.4
Carpenter (day)	0.7094A + 449.09
Glazier (day)	0.0122A + 10.31
Painter (day)	0.0905A + 37.26
Blacksmith (day)	0.479A
Beldar (day)	6.055A – 2024.87

Note : i. A is the total plinth area (m^2) of all the floors of the building.

ii. The relations are applicable for a range of 1,600 m^2 to 26,000 m^2 plinth area.

iii. The type of foundations considered is column footing.

iv. The relation for reinforcement has been given in terms of deformed bars, 10 to 20% more of this reinforcement may be taken for mild steel.

v. In case of raft foundation, additional quantities of 10 m^3 of R.C.C. 1 : 2 : 4 along with 1.2 tonnes of reinforcement per 100 m^2 of plinth area may be added.

7.3.0 OVERALL ECONOMY OF A BUILDING

The overall economy is required to be assessed for studying the economic viability of the project. The overall economy of building is generally assessed on the basis of cost of construction of the building; no cognizance is given on the cost-in-use of the structure. Cost-in-use of a building includes the capital cost of the building, its renewal cost after expiry of the estimated life, cost of maintenance and repairs of the building during the period under calculation and any other charges or taxes likely to be incurred for the structure.

If the economy of the structure is evaluated taking cost in use in consideration, the structure apparently appearing to be economical on the basis of capital cost may ultimately prove to be uneconomical.

7.4.0 ELEMENTAL COST ANALYSIS (ECA)

The manager in charge of building construction project has to assess the cost of the building prior to construction and even prior to framing of a detailed estimate. To facilitate quick assessment of cost of the building or its element (part) he can take the help of Elemental Cost Analysis.

Elemental Cost Analysis (ECA) is a systematic breakdown of cost data of different elements or parts of building. The building is divided into different elements – foundation, walls, roof, floors, reinforcement, finishes, door and window frames, water supply, electrical installations, etc. Cost of the elements are ascertained with respect to unit plinth area of the building. ECA helps in making primary comparison of cost of buildings. It attempts to isolate points of difference so that variations in specification and functions can be properly evaluated.

In normal practice of projecting the elemental cost analysis, cost of the elements are evaluated either on percentage basis of plinth area cost or on the basis of absolute cost. But cost of construction varies from time to time and from place to place. It is difficult to arrive at an acceptable standard. Central Building Research Institute (CBRI) has developed a method of elemental cost analysis in terms of consumption of basic materials and labour per unit plinth area of building. The data of basic materials consumed had been collected from a set of final bills of different construction agencies for a set of common specifications. After making adjustments, the consumption of materials and labour was worked out with the help of average materials and labour constants for each element.

The consumption of materials/labour for 10 sq.m. of plinth area for two and four storeyed residential buildings (load bearing construction) were tabulated and are given in Table 7.6. With this data, elemental cost analysis for any place at any time can be worked out from the rates of materials and labour prevailing in the particular place at the particular time.

TABLE 7.6.1

Materials and labour inputs for various elements and sub-elements for two and four-storeyed residential buildings (load bearing walls)

For 10 m^2 Plinth area

Figs. in lines 'a' are for four-storeyed dwellings. Figs. in line 'b' for two-storeyed dwellings.

	Elements		(i) Foundation and Plinth						(ii) Walling			(iii) Structural Floor		
Sl. No.	Materials or Labour	Sub-element unit	Exca-vation	Earth filling	Sand filling	Bed concrete	Brick work	Damp Proof course	Brick work	RCC in lintel chajja, beams etc.	Rein force-ment	RCC concrete work	Form work	Slab reinfor-cement
1	2	3	4	5	6	7	8	9	10	11	12	13	14	15
1.	Cement	a. tonne	—	—	—	0.09	0.08	0.005	0.3	0.133	—	0.44	—	—
		b.	—	—	—	0.13	0.09	0.01	0.3	0.122	—	0.44	—	—
2.	Sand	a. m^3	—	—	0.2	0.32	0.34	0.007	1.3	0.19	—	0.64	—	—
		b.	—	—	0.4	0.46	0.39	0.013	1.3	0.18	—	0.64	—	—
3.	Coarse Aggregate	a. m^3	—	—	—	0.61	—	0.013	—	0.34	—	1.113	—	—
		b.	—	—	—	0.87	—	0.03	—	0.31	—	1.113	—	—
4.	Steel	a. kg.	—	—	—	—	—	—	—	—	51.0	—	—	100.0
		b.	—	—	—	—	—	—	—	—	46.0	—	—	100.0

Contd.

	Elements		(i) Foundation and Plinth						(ii) Walling			(iii) Structural Floor		
Sl. No.	Materials or Labour	Sub-element unit	Exca-vation	Earth filling	Sand filling	Bed concrete	Brick work	Damp Proof course	Brick work	RCC in lintel chajja, beams etc.	Rein force-ment	RCC concrete work	Form work	Slab reinfor-cement
1	2	3	4	5	6	7	8	9	10	11	12	13	14	15
5.	Timber	a. m^3	—	—	—	—	—	—	—	0.0144	—	—	0.0304	—
		b.	—	—	—	—	—	—	—	0.0137	—	—	0.0304	—
6.	Ballies	a. m	—	—	—	—	—	—	—	1.20	—	—	3.71	—
		b.	—	—	—	—	—	—	—	1.13	—	—	3.71	—
7.	Bricks	a. no.	—	—	—	—	627	—	2371	—	—	—	—	—
		b.	—	—	—	—	716	—	2371	—	—	—	—	—
8.	Bitumen	a. kg	—	—	—	—	—	0.66	—	—	—	—	—	—
		b.	—	—	—	—	—	1.33	—	—	—	—	—	—
9.	Mason	a. day	—	—	—	0.07	0.9	0.035	5.6	0.24	—	0.93	—	—
		b.	—	—	—	0.1	1.01	0.07	5.1	0.22	—	0.93	—	—
10.	Carpenter	a. day	—	—	—	—	—	—	—	0.72	—	—	4.18	—
		b.	—	—	—	—	—	—	—	0.68	—	—	4.18	—
11.	Black-smith	a. day	—	—	—	—	—	—	—	—	0.50	—	—	0.95
		b.	—	—	—	—	—	—	—	—	0.45	—	—	0.95
12.	Painter	a. day	—	—	—	—	—	0.008	—	—	—	—	—	—
		b.	—	—	—	—	—	0.016	—	—	—	—	—	—
13.	Beldar	a. day	1.10	0.51	0.05	1.20	1.9	0.035	10.62	2.19	0.50	5.85	4.2	0.95
		b.	0.99	0.94	0.09	1.73	2.18	0.07	10.08	1.87	0.45	5.25	4.2	0.95

TABLE 7.6.2

Materials and labour inputs for various elements and sub-elements for two and four-storeyed residential buildings (load bearing walls)

	Element		iv. Stair case			v. Doors			vi. Windows			
Sl. No.	Material or labour	Sub element unit	RCC (conc. work only)	Form work	Reinforce-ment	Frame	Shutter	Holdfast	Frame	Shutter	Holdfast	Window grill
1	2	3	4	5	6	7	8	9	10	11	12	13
1.	Cement	a. tonne	0.019	—	—	—	—	0.007	—	—	0.006	—
		b.	0.019	—	—	—	—	0.007	—	—	0.006	—
2.	Sand	a. m^3	0.028	—	—	—	—	0.014	—	—	0.013	—
		b.	0.028	—	—	—	—	0.014	—	—	0.013	—
3.	Coarse Aggregate	a. m^3	0.049	—	—	—	—	0.027	—	—	0.025	—
		b.	0.049	—	—	—	—	0.027	—	—	0.025	—
4.	Steel	a. kg	—	—	7.5	—	—	2.4	—	—	2.3	22.4
		b.	—	—	7.5	—	—	2.4	—	—	2.3	22.4
5.	Timber	a. m^3	—	0.0012	—	0.038	0.052	—	0.04	0.021	—	—
		b.	—	0.0012	—	0.038	0.052	—	0.04	0.021	—	—

Contd.

	Element		iv. Stair case			v. Doors			vi. Windows			
Sl. No.	Material or labour	Sub element unit	RCC (conc. work only)	Form work	Reinforce-ment	Frame	Shutter	Holdfast	Frame	Shutter	Holdfast	Window grill
1	2	3	4	5	6	7	8	9	10	11	12	13
6.	Ballies	a. m	—	0.03	—	—	—	—	—	—	—	—
		b.	—	0.03	—	—	—	—	—	—	—	—
7.	Glass	a. m^2	—	—	—	—	—	—	—	0.47	—	—
		b.	—	—	—	—	—	—	—	0.47	—	—
8.	Mason	a. day	0.11	—	—	—	—	0.19	—	—	0.17	—
		b.	0.11	—	—	—	—	0.19	—	—	0.17	—
9.	Carpenter	a. day	—	0.12	—	0.4	2.5	—	0.42	1.23	—	0.13
		b.	—	0.12	—	0.4	2.5	—	0.42	1.23	—	0.13
10.	Black-smith	a. day	—	—	0.07	—	—	0.18	—	—	0.17	0.65
		b.	—	—	0.07	—	—	0.18	—	—	0.17	0.65
11.	Belder	a. day	0.27	0.07	0.07	0.2	0.68	0.23	0.21	0.37	0.22	0.20
		b.	0.23	0.07	0.07	0.2	0.68	0.23	0.21	0.37	0.22	0.20

TABLE 7.6.3

Materials and labour inputs for various elements and sub-elements for two and four storeyed residential buildings (load bearing walls)

Sl. No.	Element		vii. Flooring		viii. Roofing		ix. Finishes			
	Material or labour	Sub-element unit	40 mm PCC layer	Skirting	Lime conc.	Bitumen painting	15 mm plaster	12 mm plaster	White/colour wash	Painting
1	2	3	4	5	6	7	8	9	10	11
1.	Cement	a. tonne	0.13	0.025	0.014	—	0.086	0.072	—	—
		b.	0.13	0.025	0.028	—	0.092	0.077	—	—
2.	Sand	a. m^3	0.155	0.049	0.029	0.013	0.366	0.306	—	—
		b.	0.155	0.049	0.058	0.026	0.394	0.33	—	—
3.	Coarse aggregate	a. m^3	0.31	—	—	—	—	—	—	—
		b.	0.31	—	—	—	—	—	—	—
4.	Brick aggregate	a. m^3	—	—	0.21	—	—	—	—	—
		b.	—	—	0.42	—	—	—	—	—
5.	Lime	a. qtl.	—	—	0.33	—	—	—	0.153	—
		b.	—	—	0.66	—	—	—	0.176	—
6.	Surkhi	a. m^3	—	—	0.10	—	—	—	—	—
		b.	—	—	0.20	—	—	—	—	—

Contd.

	Element		vii. Flooring		viii. Roofing		ix. Finishes			
Sl. No.	Material or labour	Sub-element unit	40 mm PCC layer	Skirting	Lime conc.	Bitumen painting	15 mm plaster	12 mm plaster	White/colour wash	Painting
1	2	3	4	5	6	7	8	9	10	11
7.	Brick Tiles	a. no.	—	—	74	—	—	—	—	—
		b.	—	—	148	—	—	—	—	—
8.	Glass	a. m^3	0.46	—	—	—	—	—	—	—
		b.	0.46	—	—	—	—	—	—	—
9.	Bitumen	a. kg	—	—	—	3.55	—	—	—	—
		b.	—	—	—	7.10	—	—	—	—
10.	Primer	a. ltr	—	—	—	—	—	—	—	0.44
		b.	—	—	—	—	—	—	—	0.44
11.	Paint	a. ltr	—	—	—	—	—	—	—	0.85
		b.	—	—	—	—	—	—	—	0.85
12.	Mason	a. day	1.45	0.33	0.40	—	1.81	1.61	—	—
		b.	1.45	0.33	0.80	—	2.01	1.73	—	—
13.	Painter	a. day	—	—	—	0.03	—	—	—	0.53
		b.	—	—	—	0.06	—	—	—	0.53
14.	White-washer	a. day	—	—	—	—	—	—	0.92	—
		b.	—	—	—	—	—	—	1.06	—
15.	Beldar	a. day	2.15	0.47	2.22	0.08	2.86	2.41	0.92	0.53
		b.	2.15	0.47	4.44	0.16	3.07	2.59	1.06	0.53

TABLE 7.6.4

Materials and labour inputs for various elements and sub-elements for two and four-storeyed residential buildings (load bearing walls)

		Element (x) Water Supply and Sanitary Services			
Sl. No.	Materials or labour	Sub element unit	Fittings	Soil/Waste vent/ drainage pipes	Water Supply
1	2	3	4	5	6
1.	W.C. Seat (complete with all accessories and fittings)	no.	0.18	—	—
2.	Wash basin -do-	no.	0.18	—	—
3.	Sand cast iron pipe				
	a. 100 mm	m	—	1.04	—
	b. 50 mm	m	—	1.46	—
4.	Cement	tonne	—	0.065	—
5.	Sand	m^3	—	0.19	—
6.	Coarse aggregate	m^3	—	0.32	—
7.	Paint	ltr.	0.15	0.47	0.04
8.	Primer	ltr.	—	0.020	0.02
9.	Bricks	no.	—	38	—
10.	M.S. Holder Bat Clamp	no.	—	1.5	—
11.	Floor trap	no.	—	0.54	—
12.	G. I. Pipe 15 mm	m	—	—	2.5
13.	G. I. Pipe 20 mm and above	m	—	—	3.33
14.	Bib Cock 15 mm	no.	—	—	0.54
15.	Stop Cock 15 mm	no.	—	—	0.54
16.	Pig lead	kg.	—	4.2	—
17.	Spun yarn	kg.	—	0.61	—
18.	Lead connector 30 cm	no.	—	—	0.18
19.	Shower rose 100 mm	no.	0.18	—	—
20.	Tank (RCC) 270 ltr.	no.	—	—	0.10
21.	S.W. Pipe 100 mm – 150 mm	m	—	1.7	—
22.	Gully Trap 100 mm × 100 mm	no.	—	0.27	—
23.	Fitter	day	0.28	0.57	0.69
24.	Mason	day	0.15	0.73	—
25.	Painter	day	0.045	0.30	0.07
26.	Beldar	day	0.39	2.00	0.56

Computation based on medium size tenement (Pl. area 56 m^2) double storeyed construction.

TABLE 7.6.5

Materials and labour inputs for various elements and sub-elements for two and four-storeyed residential buildings (load bearing walls)

		Element	xi. Internal Electrification		
S1. No.	Materials / Labour	Sub-element unit	Wiring	Fittings	L. C. Boards Switches and distribution boards
1	2	3	4	5	6
1.	5 Amp. Tumbler switch (single pole)	no.	2.4	—	—
2.	5 Amp. Tumbler Switch (double pole)	no.	—	—	—
3.	15 Amp. Tumbler Switch (double pole)	no.	0.19	—	—
4.	5 Amp. socket	no.	0.47	—	—
5.	15 Amp. socket	no.	0.19	—	—
6.	PVC insulated A1. Conductor single core cable				
	a. 1.5 sq mm	m	26.3	—	—
	b. 2.5 sq mm	m	2.6	—	—
	c. 4–6 sq mm	m	5.0	—	—
7.	Conduit 19 mm (H.G.)	m	11.0	—	—
8.	Conduit 25 mm (H.G.)	m	1.33	—	—
9.	Socket for 19 mm & 25 mm conduit	no.	4.3	—	—
10.	Inspection bends 19 mm & 25 mm	no.	2.9	—	—
11.	Saddles or Stapplers	no.	30.00	—	—
12.	I.C. Box 75 × 75 × 60 mm (deep)	no.	2.9	—	—
13.	I.C. Box 200 × 150 × 100 mm (deep)	no.	5.4	—	—
14.	Phenolic laminated sheet	m^2	0.04	—	—
15.	Junction Box	no.	2.1	—	—
16.	Ceiling rose (2 Plate)	no.	2.11	—	—
17.	"L" type cast A1. anodised wall bracket brass socket (complete)	no.	—	1.14	—
18.	Circular ceiling fitting (complete)	no.	—	0.24	—
19.	Bulk Head Fitting (Complete)	no.	—	0.19	—

		Element	xi. Internal Electrification		
Sl. No.	Materials / Labour	Sub-element unit	Wiring	Fittings	L. C. Boards Switches and distribution boards
1	2	3	4	5	6
20.	15 Amp S.P. & N. Switch Fuse Unit	no.	—	—	0.47
21.	2 way M.C.D.B. 15 Amp./Way	no.	—	—	0.24
22.	Wireman	day	1.19	0.07	0.04
23.	Mason	day	2.24	—	—
24.	Painter	day	0.03	—	—
25.	Khalasi	day	3.71	0.07	0.04

Computation is based on tenement of 70 m^2 plinth area.

References :

1. Management by network analysis – Dr. Amitava Bhattacherjee
2. Industrial Engineering — R. B. Gupta
3. Management — D. C. Das
4. Modern Production Management —Elwood S. Buffa
5. Improvement of Construction Technology — A. Nagabhushan – Journal Institution of Engineers, Vol. 69, March '89
6. Management Techniques in Indian Construction Industry — A Questionnaire Survey – K. Ananthanarayanan and Dr. P. Kalyansundaram Journal Institution of Engineers, Vol. 72, Sept. '91
7. Construction Management at Site Work—Dr. A. K. Panda, Journal, Institution of Engineers, Vol. 73, July, '92
8. Advances in Building Materials and Construction
9. Handbook on Building Economics and Productivity — Central Building Research Institute, Roorkee
10. Maintenance and Repairs of Buildings — P. K. Guha

QUESTIONS

1. What are the objectives of Project Management? What are the differen methods of project management? **M. Arch, J.U. '96**
2. Write notes on
   - i. Administrative approval **M. Arch, J.U. '97**
   - ii. Technical sanction **M. Arch, J.U. '97**
   - iii. Snag list
3. What are the ingredients of contract?
4. What are the different ingredients of tender documents? **M. Arch, J. U. '95,'97**
5. What are the different types of contract? **M. Arch., J.U. '95,'96,'97**
6. Discuss comparative Merits/Demerits of different building contract systems. **M. Arch., J. U. '95,'96,'97**
7. What do you understand by Liquidated Damages ? **M. Arch., J. U. '95 ,' 96,'97**
8. What is escalation of price? What sort of provision need be made in the tender to sort out problems arising due to escalation of prices? **M. Arch., J. U. '95,'96,'97**
9. Write notes on Elemental Cost Analysis of Residential Buildings (ECA). What are the uses? **M. Arch, J.U. '95,'97**
10. Write a note on Material and Labour Requirements for buildings.

    What is the basis of study? Discuss with respect to Residential Buildings and Office Buildings. **M. Arch, J. U. '95,'97**
11. Discuss Critical Path Method of Construction Management. **M. Arch, J. U. '95; 96,'97**
12. Write notes on
    - a. Management control **M. Arch, J.U. '97**
    - b. Quality control
    - c. Financial Control **M. Arch, J. U. '95,'97**
13. What is the significance of Float in a network?

    Explain with example Total Float, Free Float, and Independent Float. **M. Arch, J. U. '95**
14. What is Crashing of Programme? Why Crashing of Programme is done? **M. Arch, J. U. '95,'97**
15. What do you understand by the term probability of completion of a Project? **M. Arch, J. U. '95,'97**
16. Describe NETWORK with an example of nine activities only. Tabulate EST, LST, EFT, LFT, TF, FF, IF. Show supporting calculations. **M. Arch, J. U. '94,'95**

17. What are meant by Optimistic Estimate, Most likely Estimate and Expected Time in the PERT? **M. Arch, J. U. '95**

18. Explain the Forward Pass Method and Backward Pass Method used in Network Technique. **M. Arch, J. U. '95**

19. Find out Optimum Time required for completing construction of your own building by a Bar-Chart diagram. **M. Arch, J. U. '94**

    You may use the data given in Annexure II of a Building having a covered area of 90 sqm (single-storey building). **M. Arch. J.U. '94**

20. Discuss Integrated Management for Building Project. **M. Arch, J. U. '95, '96,'97**

21. Write on
    a. Planning
    b. Monitoring
    c. Scheduling
    d. Control **M. Arch, J. U. '94**

22. What are the advantages of Line of Balance Technique in Construction Management? **M. Arch, J. U. '94**

23. With reference to an example describe application of Line of Balance with graphical representation of activities and units of production. **M. Arch, J. U. '94**

24. Describe the relation between L.O.B. and NETWORK technique for repetitive activities of a construction management program.

    Write on manpower planning. **M. Arch, J. U. '94**

25. What is Master METWORK? When a Master Network becomes necessary?

26. Write a note on the specifications adopted both for load bearing and framed construction of Residential buildings – on elements like – Foundation, Frame, Walling, Floor, Roof, Joining, Finishes, Stairs, Fittings and Fixtures. **M. Arch, J. U. '95**

27. What are the advantages and limitations of Bar-Chart?

    What are the differences between Bar-Chart and Network?

    M. Arch, J. U. '95,'96

28. What are the advantages of Line of Balance Technique? Describe how it is drawn. **M. Arch, J. U. '96**

29. Write short notes on
    a. Floats in Network **M. Arch, J.U. '97**
    b. Financial Control
    c. Crashing of Programme
    d. Master network **M. Arch, J.U. '97**
    e. Probability of completion of a project
    f. Percentage progress. **M. Arch, J. U. '96**
    g. Budget allotment **M. Arch, J.U. '97**

Note : J. U. means Jadavpur University.

ANNEXURE I

CHECK/SNAG LIST

Name of the Project ....

CHECK LIST

House No. is ready in all respects to receive Render – State Yes/No. ....

Total Man-hours used :

Area		Item	Yes/No	Comments
PARAPET WALL	1	All rendered brickwork is sealed and ready for render.		
	2	All fairfaced brickwork is in an acceptable condition.		
	3	In situ concrete banding is fair faced.		
	4	Mortar fillet to base of wall is complete.		
	5	No Bricks are missing.		
	6	Claustra 75 mm is the correct type and workman like		
	7	Is the in situ concrete of the spout fair faced both sides of the parapet wall?		
	8	Is the groove for felt completed in an acceptable fasion?		
	9	Vent pipe finished off correctly.		
ROOF SURFACE	1	Is the Roof clean ?		
	2	Is the finish acceptable?		
	3	Is the in situ step in place, in an acceptable condition?		
STAIRWAY	1	All concrete snots/projections have been removed.		

Area		Item	Yes/No	Comments
	2	Concrete to be rendered has been fitted with autex.		
	3	Brickwork to the stair is complete in all respects.		
	4	The stairs & landings are clean.		
	5	The outlet pipe to bottom landing is clean & protected.		
ENTRANCE	1	All exposed concrete to be rendered has autex fitted.		
	2	Infilling of conduits and electrical fittings is complete and correct.		
	3	The ceiling is free from lips.		
	4	The area is clean.		
LOBBY	1	The D. B. and its associated conduits have been infilled correctly.		
	2	The block is complete in all respects.		
	3	The ceiling is free from lipping.		
BEDROOM No. 1	1	Infilling to electrical conduits and fittings is correct.		
	2	Sills to windows have been fitted correctly.		
	3	The ceiling is free from lipping.		
	4	The room is clean and tidy.		
BED ROOM NO. 2	1	Infills to all conduits and electrical outlets are complete and correct.		
	2	Sills to windows have been fiftted correctly.		
	3	Infilling of plumbing is complete and correct.		
	4	The ceiling is free from lipping.		

Area		Item	Yes/No	Comments
	5	The room is clean and tidy.		
DRAWING ROOM	1	Infilling to all conduits and electrical fittings is complete and correct.		
	2	Window sills have been fitted correctly.		
	3	The room is clean and tidy.		
KITCHEN	1	Infills to plumbing are complete and correct.		
	2	Infills to electrical conduits & outlets are complete and correct.		
	3	Kitchen shelf is in an acceptable condition.		
	4	The fan outlet has been shaped correctly.		
	5	The ceiling is free from lipping.		
	6	The room is clean and tidy.		
BATHROOM	1	Infills to electrical conduits and outlets are complete and correct.		
	2	Infilling of all plumbing is complete and correct.		
	3	Brickwork to the window has been sealed.		
	4	The ceiling is free from lipping.		
	5	The room is clean and tidy.		
	6	The recesses for the squat-pan and shower are clean.		
TOILET	1	All plumbing infilling is complete and correct.		
	2	Brickwork to the window is sealed.		
	3	Brickwork to the doors is sealed.		

Area		Item	Yes/No	Comments
	4	The ceiling is free from lipping.		
	5	The floor recess for squat-pan is clean.		
	6	The room is clean and tidy.		
GENERAL	1	Ceiling fan hook recesses shaped correctly.		
EXTERNAL	1	All claustra has been fitted		
	2	All claustra is workmanlike		
	3	All exposed concrete to be rendered has been fitted with autex.		
	4	All concrete projections/snots likely to interfere with render have been removed.		
	5	The washing trough is complete, i.e. plumbing mortared in, and the brick-work complete and correct in all respects.		
	6	The fairfaced part of the stair landing is actually fairfaced.		
	7	There is no hole in the brickwork, i.e. brickwork is complete.		
	8	All electrical outlets have been mortared in and the services protected.		
	9	All plumbing outlet infills are complete and correct.		
	10	No blocks or claustra lying around. They have been removed and stacked with main stock.		
	11	No scaffolding is lying around (it has been removed and is in sorting area)		
SIGNED BY	1	**Engineer**		
	2	**Foreman**		
	3	**Charge-hand**		

ANNEXURE II
CHART OF MAN-DAYS

SL NO.		NAME OF THE ITEM OF WORKS	QUANTITIES	WORKS BEFORE THIS ITEM OF WORKS	WORKS DURING THIS ITEM OF WORKS	WORKS THAT FOLLOW THIS ITEM OF WORKS	TOTAL MAN-DAYS	NO. OF LABOURS OR MASON	NO. OF TOTAL DAYS	REMARKS
1.	1(a)	Earth excavation	45.34 cu.m		2(a)	3(a) 4(a)	10	5	2	
	1(b)	Earth filling in foundation	16.8 cu.m	4(a)	1(d)	5	5	5	1	By providing overtime.
	1(c)	Earth filling in plinth	25.2 cu.m	5	8(a)	4(b)	5	5	1	
	1(d)	Earth ramming	12.5 cu.m	8(a) 4(b)	2(b)	3(b)	15	5	3	
2.	2(a)	Flat brick soling in foundation	47.7 sq. m	3(a) 3(c)	4(a) 4(d)	4(a) 4(d)	3	3	1	
	2(b)	Flat brick soling in plinth	41.3 sq.m	4(b) 8(a)	6(a)	3(b) 3(c)	3	3	1	By providing overtime.
	2(c)	Flat brick soling in underground tank	1.09 cu.m	3(c)	4(d)	14(a)				
3.	3(a)	P.C.C. in foundation	7.11 cu.m	4(a)	5	1(b)	6	6	1	
	3(b)	P.C.C. in plinth	6.19 cu.m	2(b)	5	4(b) 8(a) 1(d)	6	6	1	By providing overtime.
	3(c)	P.C.C. in underground tank	0.54 cu.m	2(b)	4(d)	14(a)				

Contd.

SL NO.		NAME OF THE ITEM OF WORKS	QUANTITIES		WORKS BEFORE THIS ITEM OF WORKS	WORKS DURING THIS ITEM OF WORKS	WORKS THAT FOLLOW THIS ITEM OF WORKS	TOTAL MAN-DAYS	NO. OF LABOURS OR MASON	NO. OF TOTAL DAYS	REMARKS
4.	4(a)	Brickwork in foundation up to plinth	18.1	cu.m	5, 8(a)	1(b)	1(d) 2(b)	28	7	4	
	4(b)	Brickwork up to below lintel level	17.53	cu.m	5	8(a)	6(a)	35	7	5	
	4(c)	Brickwork below roof level.	8.55	cu.m	6(a)	6(b)	6(b)	18	6	3	
	4(d)	Brickwork at underground tank	8.16	cu.m	1(a) 2(a)	14(a)	1(b)	8	4	2	
	4(e)	Brickwork at parapet wall	6.3	cu.m	6(b) 6(c)	6(d)	7	16	4	4	
	4(f)	Brickwork at overhead tank	0.75	cu.m	4(e)	14(b)	14(b) 6(d)	4	2	2	

CHART OF MAN-DAYS

SL NO.		NAME OF THE ITEM OF WORKS	QUANTITIES	WORKS BEFORE THIS ITEM OF WORKS	WORKS DURING THIS ITEM OF WORKS	WORKS THAT FOLLOWS THIS ITEM OF WORKS	TOTAL MAN-DAYS	NO. OF LABOURS OR MASON	NO. OF TOTAL DAYS	REMARKS
5.	5.	D.P.C. at plinth level	3.70 cu.m	4(a)	8(a)	1(b), 1(d) 2(b)	4	4	1	
6.	6(a)	R.C.C. at lintel with chajja	1.71 cu.m	4(b)	4(c)	6(b)	12	6	2	By Providing overtime
	6(b)	Preparation of shuttering and Rod binding	5.5 sq.m	4(c)	6(c)	7.4(e)	30	6	5	By Providing overtime
	6(c)	R.C.C. roof casting	5.1 cu.m	6(b)	4(c) 7	4(f)	16	16	1	
	6(d)	R.C.C. roof underground & overhead tank	0.69 cu. m	14(a)	1(b)	13(a)	4	2	2	
7.	7.	Roof protection (lime terracing)	52.27 sq.m	4(e) 4(f)	14(b)	11(d)	112	8	14	
8.	8(a)	Woodwork in door & window frame Preparation & Fixing	0.14 cu. M	5, 4(a)	4(b)	4(c)	4	2	2	

Contd.

SL NO.		NAME OF THE ITEM OF WORKS	QUANTITIES		WORKS BEFORE THIS ITEM OF WORKS	WORKS DURING THIS ITEM OF WORKS	WORKS THAT FOLLOWS THIS ITEM OF WORKS	TOTAL MAN-DAYS	NO. OF LABOURS OR MASON	NO. OF TOTAL DAYS	REMARKS
	8(b)	Woodwork in door & Window Shutters	1.03	cu.m	12(b) 12 (a)	9(a) 9(b)	17(a)	28	2	14	
9.	9(a)	Grill in windows	7.35	cq.m	12(b)	8(a) 8(b)	17(a)	9	3	3	
	9(b)	Grill doors	3.57	sq.m	12(a)	11(b) 11(c)	17(a)	2	2	1	
10.	10(a)	Chipping at lintel & chajja	10.5	sq.m	7, 4(e)	11(b) 11(c)	18(a)	3	3	1	
	10(b)	Chipping at roof	51.0	sq.m	7, 11(b)	11 (e)	7	6	3	2	
11.	11(a)	Plaster at ceiling	41.5	sq.m	10(b)	11(b)	11(c)	10	5	2	By Providing overtime
	11(b)	Plaster at inside wall	177	sq.m	18(a)	11(c) 12(a)	18(b) 15(a)	40	8	5	
	11(c)	Plaster at outside wall & chajja	118	sq.m	10(a)	13(a)	16(a)	30	6	5	
	11(d)	Plaster at overhead tank	10.14	sq.m	4(e)	4(f) 14(b)	7	2	2	1	By Providing overtime
	11(e)	Plaster at parapet inside	25.5	sq.m	4(e)	7	11(b) 11(c)	6	3	2	

CHART OF MAN-DAYS

SL NO.		NAME OF THE ITEM OF WORKS	QUANTITIES	WORKS BEFORE THIS ITEM OF WORKS	WORKS DURING THIS ITEM OF WORKS	WORKS THAT FOLLOWS THIS ITEM OF WORKS	TOTAL MAN-DAYS	NO. OF LABOURS OR MASON	NO. OF TOTAL DAYS	REMARKS
12.	12(a)	Ground Floor flooring (I.P.S.)	41.55 sq.m	11(a), 11(b) 13(a)	9(a), 13(b)	15(a) 17(a)	15	5	3	
	12(b)	Telanchi at window	3.21 sq.m	9(a), 8(b)	8(b)	18(b) 15(a)	2	2	1	
	12(c)	Dado of toilet	16.8 sq.m	11(b), 13(b)	13(b)	15(a)	4	2	2	
13.	13(a)	Toilet & Kitchen C.I., G.I. Pipe Laying and tank		11(b), 11(d) 11(c)	12(a)	15(a) 16(a)	6	3	2	
	13(b)	Toilet fittings & kitchen fittings		12(a), 15(a) 18(a), 18(b)			2	2	1	
14.	14(a)	I.P.S. inside under-ground tank.	8.00 sq.m	4(d), 9(d)	13(a)	11(d)	2	2	1	
	14(b)	I.P.S. at overhead tank (inside) preparation & fixing	11.4 sq.m	4(f), 6(d)	13(a)	11(c)	2	2	1	By providing overtime

Contd.

SL NO.		NAME OF THE ITEM OF WORKS	QUANTITIES	WORKS BEFORE THIS ITEM OF WORKS	WORKS DURING THIS ITEM OF WORKS	WORKS THAT FOLLOWS THIS ITEM OF WORKS	TOTAL MAN-DAYS	NO. OF LABOURS OR MASON	NO. OF TOTAL DAYS	REMARKS
15.	15(a)	White wash (inside)	177 sq.m	11(a), 11(b) 13(a)	13(b)	17(a)	6	2	3	
16.	16(a)	Outside colour wash	118 sq.m	11(c), 13(a)	13(b)	18(b)	6	2	3	By providing overtime
17	17(a)	Synthetic paint at door & window and grill	52.4 sq.m	8(b), 9(a) 9(b)	13(b)	18(b)	6	2	3	
18.	18(a)	Electrical wiring pipe laying		4(c), 6(c)	11(b)	11(c), 11(b)	4	2	2	
	18(b)	Wiring D. B. Box Pump sets etc.		18(a), 15(a) 16(a)			6	3	2	By providing overtime